The Millionaires' Squadron

The Millionaires' Squadron

The Remarkable Story of 601 Squadron and the Flying Sword

Tom Moulson

Pen & Sword
AVIATION

First published in Great Britain in 2014 by
Pen & Sword Aviation
an imprint of
Pen & Sword Books Ltd
47 Church Street
Barnsley
South Yorkshire
S70 2AS

ISBN 978 1 78346 339 8

A CIP catalogue record for this book is available from the British Library

Typeset in Ehrhardt by
Mac Style, Bridlington, East Yorkshire
Printed and bound in the UK by CPI Group (UK) Ltd, Croydon, CRO 4YY

Pen & Sword Books Ltd incorporates the imprints of Pen & Sword Archaeology, Atlas, Aviation, Battleground, Discovery, Family History, History, Maritime, Military, Naval, Politics, Railways, Select, Transport, True Crime, and Fiction, Frontline Books, Leo Cooper, Praetorian Press, Seaforth Publishing and Wharncliffe.

For a complete list of Pen & Sword titles please contact
PEN & SWORD BOOKS LIMITED
47 Church Street, Barnsley, South Yorkshire, S70 2AS, England
E-mail: enquiries@pen-and-sword.co.uk
Website: www.pen-and-sword.co.uk

Contents

Acknowledgements

Several quotations are made in the text and the publishers wish to thank the following for permission to use them: the Controller of Her Majesty's Stationery Office, for permission to quote Crown Copyright material in the form of extracts from 601 Squadron's official diary and combat reports; Bill Matheson for permission to quote from *A Mallee Kid With The Flying Sword*, published by W. M. Matheson; *Flypast* magazine for permission to quote from 'Desert Encounter' about Ray Sherk; Penguin Books for permission to quote from Geoffrey Wellum's book *First Light*, and The United States Naval Institute Press for permission to quote from John Frane Turner's book, *Fight for the Air*. Quotations are also made and acknowledged from sources that can no longer be traced: William Heinemann Ltd for the second volume of *Assignment to Catastrophe* by Major-General Sir Edward Spears; and Hutchinson & Co. for *Gran Chaco Adventure* by T. Wewege Smith.

The author is indebted to all those who helped bring this book's predecessor, *The Flying Sword* (1964) to fruition. Fred Tomlinson, formerly news editor of *The Observer*, graciously gave the author a wealth of notes on the pre-war era, which he had gathered for an abandoned book. Brian Thynne, a polished raconteur, in his Sussex mansion and over several enjoyable lunches in London gave me a clear sense of the squadron's brand of professionalism and mischief in the pre-war days. Sir Archibald Hope in his director's office at Napiers was able to recall vividly details of the Battle of Britain with names, dates and anecdotes. Sir Max Aitken in his chairman's office at the Beaverbrook Press in Fleet Street furnished much of the story until the end of the Battle of Britain, and generously lent me his log book. Whitney Straight, at the time Deputy Chairman of Rolls-Royce, was at first unwilling to turn back the clock, but after relenting couldn't possibly have been more helpful, talking freely and lending me his contemporaneous diary of events from Norway to France, captivity, and escape. John Bisdee, who took weeks to track down, was occupying the chairman's office of a Unilever company in London's West End, two hundred yards away from his old

Intelligence Officer, Ken Carew-Gibbs, whom he hadn't seen since North Africa, and who was equally unaware of this proximity; he also lent me his log book. Sir Dermot Boyle; Viscount (George) Ward; John Parkes; Aidan Crawley and his wife Virginia Cowles; Lady Perkins, widow of Sir Nigel Norman; Flight Lieutenant Marshall; Loel Guinness, and many others gave invaluable help. All these people impressed me with their openness, modesty, and obvious affection for what we called the Legion, and the only reason they would give so such time to a novice author was that he had been a Legionnaire himself.

More recently, sincere thanks for helping with this updated, enlarged, and renamed edition are extended to Brian Thynne's daughters Georgina, Harriet, and Ulrica Thynne and Penelope Charteris for Thynne's fascinating unpublished manuscript, *My Years In 601*; to Paul Doulton for his recollections and his mother Carol's moving letter to her father in America during the Battle of Britain; to Billy Clyde's daughters June, Alicia, and Gail for their mother Barbro's letters; to Paul Harnden for permission to quote from his grandfather Flight Sergeant Gilbert Henry Harnden's diary; to Harold Brown's family and Trevor Hipperson for permission to quote from Harold Brown's diary; and to my long-time squadron 'oppo' Denis Shrosbree for his post-war recollections, or perhaps confessions. Jennifer Schwartz in America, who is conducting research on Roger Bushell, put me right on several points in my draft. As one of the longest survivors of 'The Few', Jack Riddle contributed oral and written reminiscences of life before and during the Battle of Britain, in the air and on the ground, shortly before he died in 2009.

Far from least, three people must be mentioned without whose urging this updated and enlarged edition of the former book would not even have been attempted, and without whose research efforts it would not have been possible. John Wheeler and Jamie Ivers, Americans who became interested in the story of 601 Squadron, created a website around it which served as a magnet for relatives of the squadron's past members and which contains a marvellous collection of photographs.

Much of their material originated with Ed McManus, an aviation enthusiast and private pilot in the UK who served on the committee for the Battle of Britain Monument unveiled on London's Victoria Embankment in 2005. Ed's responsibility for gathering 2,937 names to be engraved on the memorial put him into touch with members of 601 worldwide still alive and with the families of those who no longer are.

Sincere acknowledgments and thanks are extended to Laura Hirst, Pen and Sword Books Aviation Imprint Administrator, who coordinated the manuscript, photographs, jacket design and editing and shepherded the project through to finality; to Jon Wilkinson, the jacket designer, who took note of my ideas; and to Ting Baker, editor, who patiently fielded countless running changes. Jacket photograph by courtesy of the RAF Museum.

NOTE: Two paragraphs in Chapter 11 about Roger Bushell are also to be found in Simon Pearson's excellent 2013 book, *The Great Escaper*. In fact they were quoted from this book's 1964 predecessor, *The Flying Sword*, with attribution to Tom Moulson mistakenly omitted.

Preface

So This is How it is!

It's better to be on the ground wishing you were in the air than in the air wishing you were on the ground.

Aviation proverb

On the Sunday afternoon of 23 March 1952 the south of England lay moist under drizzle and cloud packed from eight hundred up to twenty thousand feet. Six DH Vampires sat at the end of runway at North Weald, Essex, their jet engines connected to external battery trolleys. The pilots were on standby for an air defence exercise and, miserable though the day was, for one of them it would soon become a lot more so. Blips to the east appeared on the controllers' green radar screens and the message to scramble came through. At a sign from the flight leader the ground crew switched on the battery accumulators. The engines came to life in a rising polyphonic whine, ground crew pulled the chocks away, and our aircraft rolled onto the runway in pairs, gathered speed while straightening out for take-off. I was number two, eyes fixed on the leader as we lifted from the runway, raised our undercarriages, and were quickly swallowed by the overcast, breaking cloud into brilliant sunshine twenty minutes later, some fifty miles out over the North Sea. The controller directed us onto a formation of incoming Dutch twin jet Gloster Meteors and we were just beginning our gun-camera attacks when my Vampire's nose dropped abruptly and all sound stopped. My body almost exploded with instant decompression. I looked with disbelief at the rev counter, rapidly unwinding to zero. Sure enough, I had flamed out.

You couldn't relight a Vampire's jet engine. I turned towards the coast and slipped back into the dense clouds. With difficulty I called 'Mayday', my voice but a hoarse croak in the rarified air. The controllers gave me a south-westerly course, which I tried to hold, but the gyro instruments began to topple. The oxygen still flowed and airspeed and altitude indicators weren't affected, but loss of the gyro instruments in this murk would outwit my primitive animal senses and spell disorientation and the famous death spiral. Stay calm. Think. Think about what? About your parents and your bedroom back home? Any flashbacks? No. Prayers? No. Regrets? Of course, I wish I had some other bloody Vampire. Have a plan. Like what? Suppose I don't spin, then, ditch? The booms will break off and kill me or pitch this useless capsule into the frigid March water with me in it. Get out now then. How? No ejector seat. No way of baling out of a twin-boom Vampire without being sliced by the tailplane. But wait, couldn't you undo your straps, eject the canopy, roll on your back, and kick the stick forward? What, am I mad – in cloud?

There's nothing I can do. This isn't another plain old brush with death but the profound, numbing, crystal-clear certainty a person can only know once, signalled by the bouncing of my legs on the rudder pedals, that the angel of death has spread its wings and my body knows it and there's nothing I can do to stop it. This rotten plane, the clouds, and the North Sea have blocked all escape. Neither I nor my comrades nor Air-Sea Rescue nor the entire Royal Air Force can do a thing to help. I shall soon die; I just don't know how, and possibly never will. They'll say, 'at least he died doing something he loved', and stuff like that. It's happened to others, now it's my turn.

So this is how it is!

Then I discovered there was something I could do after all. The cloud blended from slightly lighter above to darker below. It wasn't a horizon but it might do. Now my instrument drill was airspeed and sky, airspeed and sky, and by keeping it that way I managed to hold the wings about level and course roughly constant while gliding, in deathly silence, for the longest fifteen minutes of my life, hoping they would end over dry land. The cloud fragmented into wisps at six thousand feet with the grey waters of the English Channel below and the chalk cliffs of Folkestone a few miles to the right. How beautiful they looked! I was happy now: it would only be a crash-landing. Crossing the cliffs and spotting a grass airfield within gliding range which later turned out to be Hawkinge – by no means was I map reading – I went straight in with wheels up. Like most pilots making a dead-stick landing, to avoid overshooting I made the opposite mistake, came in too fast, and had to force the plane onto the ground before it was ready. The belly struck twice, jolting and slewing, skidding sideways, and leaving a trail of wreckage behind. The noise was deafening and it seemed to take an eternity for everything to stop. Though my harness was extremely tight I must have been thrust violently forward, my head thumping the gun sight, but I was too relieved to notice. Although there was no fire, old B for Baker was now obviously a Category Four, but I had no sympathy for it. A 'meat wagon' arrived quickly, this being a staffed WAAF base. Seeing blood on my forehead the doctor pressed me to go to the sick bay, but I declined because I felt fine. I really did. He warned me there would be a reaction, but there never was and I was elated for weeks. I was very, very happy to be alive, and still am.

The experience gave me a rare insight into the fate of many in this book who died in their cockpits after a prolonged period of terror with the end in clear sight but from which there was no escape and about which nobody would ever know, and whose official epitaphs could only be 'failed to return', 'lost on a mission', 'posted missing', and 'not seen again'. In a war my mishap would be of no consequence, just a close call. In any case I was okay, so no big deal. But it would be more than a passing event in the average civilian's week, and next morning, being a Monday, I was back as usual at my job selling telephones in London.

Because I wasn't an average civilian, but a pilot in the Royal Auxiliary Air Force, not a mere reserve but a combat-ready part of Fighter Command, with its own aircraft, ground staff and pilots on a regular RAF base. My unit was Number 601 (County of London) Squadron. It was not only the first Auxiliary unit, but the most remarkable.

T. J. M.
Newport Beach, California
January 2014

Chapter 1

The Spirit of Grosvenor

We'll surprise you yet!

Lord Edward Grosvenor, 1924

The east side of St James's Street, Mayfair, is lined with handsome Edwardian buildings whose facades have barely changed since the nineteenth century when they were the town houses or homes of English nobility. James Lock & Co., the bespoke hatter, towards the south end, still possesses the design hand-drawn by Horatio Nelson for his famous admiral's hat, and two doors further down is John Lobb the Bootmaker, which can show you the lasts made for the feet of Queen Victoria and King George V. Near the north end and close to Piccadilly stands No. 37, White's Club, or simply White's. A former gambling den where two members once bet £3,000, a fourtune in the early nineteenth century, on which of two raindrops would reach the bottom of a window first, it is still what it always was: an exclusive club for very rich gentlemen who make the selection cut, part of that cut being staunchly Tory, or today Conservative. The most exclusive gentleman's club in London, White's members have included The Prince of Wales, later Edward VII; Lord Rothschild the banker; Evelyn Waugh the writer; Oswald Mosley the British fascist; and Beau Brummell the fop. They also included Lord Edward Arthur Grosvenor, creator of No. 601 Squadron and known by his friends as 'Ned'.

Grosvenor was an eccentric, but a lovable one. He would mount the steps to the club briskly, hand his silver-topped cane, a symbol of class and authority, to the porter and announce his arrival with the call 'Hick-boo', which nobody ever seemed to understand. He would settle into a seat, order a brandy, and talk. He would talk about aeroplanes. About flying. About a phenomenon that would have been incomprehensible to everyone a generation ago.

Grosvenor was born in 1892, the son of the first Duke of Westminster, with a family name redolent of London streets, squares, and parks his ancestors had owned and developed. He was a physically impressive man of thirty-two, six-feet-two and strongly built, powerful in personality. From the outset of his life Grosvenor was an adventurer. After leaving Eton he served in the French Foreign Legion, probably a unique distinction for a member of the English aristocracy. When the excitement he sought eluded him in the desert of North Africa he bought himself out of the Legion and returned to England where, for perhaps the last time in his life, he followed the convention of his class by joining the Royal Horse Guards.

But Grosvenor felt something was missing. When he was nine years old an event three thousand miles away had stunned the world and thrilled young Ned. On a windswept beach on the outer banks of North Carolina's Atlantic coast the Wright brothers Orville and Wilbur succeeded in demonstrating the possibility of powered, controlled flight, in an ungainly biplane design with a horizontal stabiliser at the front and two pusher propellers behind the prone pilot, connected by chain drive to the engine, and with skids in place of wheels. Already balloons had carried people up and over great distances, but they were not flown; they couldn't be tacked against the wind or turned at the direction of the pilot. The Wright brothers proved that a heavier-than-air machine could be lifted from the ground and controlled through all three axes. This was incredible enough. But a mere six years later and closer to home the French aviator Louis Blériot stirred as much alarm as admiration when he crossed the English Channel from Calais to Dover in an aeroplane of his own design, winning a *Daily Mail* prize of £1,000. Now, it dawned on Britain, she was vulnerable to attack and even invasion by continental enemies. What had been dismissed as nightmare fantasy was now real, the possible would quickly become normal, and in a very few years the military value of aeroplanes would be beyond question.

Grosvenor realised this threat at the same time as he felt the siren call of aviation. Blériot was made, and as soon as he was able to turn out his monoplanes in any number Grosvenor bought two of them and learned to fly at Brooklands race track at Weybridge, Surrey, obtaining Certificate No. 607 'without', as the newspapers said, 'breaking even a wire'.

Like many others Grosvenor was intoxicated by the miracle of flight in an open-cockpit stick and rudder aeroplane. He thrilled at the roar of the engine at full throttle, the sight of the ground falling away and the horizon unrolling infinitely, the fantastic downward view of earth and the tops of clouds, the exhilaration of diving and swooping like a swallow, of descending to a satisfying rumble of wheels on grass and the engine's slowing to a 'tecka-tecka-tecka', a wondrous experience unimaginable throughout human history until now, all bursting into reality in his own lifetime.

He also saw beyond the fun. He remembered Blériot and what that meant. When the First World War came he gave one of his planes to the British government and flew the other to France where he offered it with himself to the Royal Naval Air Service. They taught him what they knew about military flying, which was little enough, and Grosvenor continued with his self tuition. *Flight* magazine reported with awe that 'at the Blériot School at Buc, on Thursday of last week, Lord Edward Grosvenor succeeded in looping the loop, while on the following day Mr Skene also accomplished the feat. Both pilots made similar flights on Saturday'.

An experience while Grosvenor was flight commander in the RNAS that he enjoyed embellishing illustrates the primitive state of aviation at the time. He briefed a new pilot at Eastchurch on the south shore of the Thames Estuary, to report to his new squadron at Ostend, Belgium. The pilot could take his aircraft off and land it again in still air, but that was about the limit of his skill; he could neither read a map nor follow a compass course. Grosvenor waited for a perfectly clear, sunny day without a breath of wind, and took the fledgling by the shoulder, led him to a corner of the hangar and pointed to the east-facing Kent coast. 'Take off that way,' he advised, 'that's east, and when you reach the sea turn right. That's south. When you get to France turn left, that's east again, and Ostend is a little way up on your right'. The pilot walked to his plane muttering softly. An hour later, Grosvenor was called to the telephone.

'I'm sorry', a voice began, 'but I must have got one of those turnings wrong and now I'm in a hell of a mess. Could somebody please come and pick up my aeroplane?'

'Where is it?' asked Grosvenor.

'Ah, it seems I'm in the middle of, ah, the tramlines at Hammersmith Broadway …'

When the war ended Grosvenor helped popularise private aviation, which he thought the government was wrong in neglecting to do, and his benevolent figure in suit and cloth cap becoming a familiar sight at air races at Lympne and Croydon aerodromes. In 1923 he endowed the Grosvenor Challenge Cup, a handsome creation of the Goldsmiths and Silversmiths Company, accompanied annually by a first and second prize of a £100 and £50. Competing aircraft, engines and pilots had to be British, and companies couldn't enter.

But all the while a larger idea was forming in Grosvenor's mind. As he assembled around him at White's enthusiastic friends with whom he could converse about his wartime flying days, he would ponder, why shouldn't there be a part-time air force? Loving aeroplanes more than horses, he visualised such a force as successor of the old mounted yeomanry, an elite corps composed of the kind of young men who in the past would have been interested in horses but were now attracted to flying machines, who had leisure and a thirst for adventure, and were prepared to indulge it in the service. The first obstacle, however, was that there wasn't as yet even an air force.

In 1914 and throughout the First World War Britain's air power was divided between the army's Royal Flying Corps (RFC) and the Royal Naval Air Service (RNAS). After the war the idea of consolidating the two forces to reduce costs was being studied by the government, the very idea of which left the army and the navy aghast. Both spluttered that the idea of an independent air force was a fantasy, that no fighting force of any quality could be created from scratch. Senior officers of both forces looked at each other and wondered how mere operators of flying machines could be as smart and officer-like as they. But just when it seemed they would prevail and air power would remain divided, the Secretary of State for Air, Sir Samuel Hoare, pressed for a full enquiry. As a result, and largely for economic reasons, in 1918 control of air power was prised from its parents and consolidated as an independent Royal Air Force (RAF). Over this was established an Air Ministry with Lord Hugh ('Boom') Trenchard as its first Chief of the Air Staff (CAS).

Forever known affectionately as father of the Royal Air Force, Trenchard, a wartime RFC pilot, overcame relentless efforts by the navy and the army to dismember the new force. He went further. In a 1919 memorandum he argued for a home defence air force of both regular and non-regular squadrons, roughly two-thirds devoted to defence. Hoare, when later Lord Templetion, recalled in his work, *Empire of the Air*: 'Trenchard realised that a fighting service must possess a non-regular branch with its roots firmly set in the civil life of the country.'

This was exactly what Grosvenor was asking for. He was ahead of the times, but not by much. It would take circumstances and politicians. The circumstances were the dire economic straits Britain was in after an exhausting war and with a far-flung empire to defend. The politicians were Hoare and Prime Minister Bonar Law. Neither was an air expert, but Trenchard was, and he had a respect for civilians as potential officers and artisans that seems unremarkable today. He had been impressed when after mistaking a fitter, newly inducted to the service, for a rigger, the fitter had nevertheless rigged an aircraft faultlessly by using a blueprint and common sense. He was convinced that there was 'material' in civil life which the air force couldn't afford to ignore.

The citizen air force, Trenchard proposed, should be split into the Special Reserve Air Force, corresponding to the militia, and the Auxiliary or Territorial Air Force. The latter were not to be used as wastage in war, but kept in squadrons, each with a companionable mess and a distinctive life of its own – in effect its officers and men should be given the life and influence of a good regiment. Sensing the wind direction, Grosvenor hoisted his sails to catch it. He gathered his chosen few for dinner at White's. 'It will happen', he assured them, 'so let us together form the first squadron of the territorial air force,' and it went without saying that he would command it. True, he admitted, no legislation had yet been passed; nor were there any administration, airfield, number, badge, or adoptive county, but these would all follow; the spirit of the squadron was in being. This, for all believers, was the moment of creation.

The RAF's professional airmen looked down on the concept of a part-time air force with as much disdain as the army and navy had looked down on them. They puffed that it would be dangerous to allow purely weekend flyers to handle the supposedly complex service aircraft of the day, that the step from horses or lorries to flying machines was too great.

An election drove the Conservative Party from office and it was the short-lived Labour government of 1924 that was responsible for passing the Auxiliary Bill into law. Another election returned Hoare in 1924 and restored to him the welcome responsibility of beginning the force. Grosvenor was ebullient. Although his squadron would later embrace the happy legend that Grosvenor singly conceived and delivered the Auxiliary Air Force, obviously he didn't. But his influence was significant. He had offered and then provided a ready-made flying unit of pilots and a commanding officer. Grosvenor never doubted that he was the Auxiliaries' father. In his own words he had 'sat on the doorstep of "Sammy" Hoare and "Boom" Trenchard' until he had his way.

Hoare expressed his own delight by becoming the first honorary air commodore of Grosvenor's squadron, thus giving his personal endorsement to a force which, despite initial doubts, eventually became the right hand of the regular air force. 'The experiment', he later recorded in *Empire of the Air*, 'was successful from the beginning. So far from the non-regular units damaging the reputation of the regular squadrons they added some of the most glorious pages to the history of the Royal Air Force during the second war.'

Five squadrons were set up on a territorial basis: two in London, one in Birmingham and two in Scotland. Grosvenor took command of the 601 (County of London) unit. The county is the sprawling area we know of as London, the city of London being a tiny area within it. At the beginning there was something of an Auxiliary honeymoon as the squadrons drew together in support of their new force. Grosvenor's reply to the still vocal critics was an angry snort and the words, 'Just wait – we'll surprise you yet!'

No. 601 Squadron was gazetted on 14 October 1925 and opened for business at Northolt with Grosvenor, two regular officers and twenty-one airmen. The aircraft were war-era Avro 504s, the squadron's rôle bombing. Grosvenor's selection criteria for officers were subtle. Owning one's own aeroplane and knowing how to fly would be a plus, being rich was even more so, but hailing from the right social class was a requirement. That was the subtle part. The first commanders of the two flights were Robin Grosvenor, Ned's nephew, and Bill Collett, son of the Lord Mayor of London. Nepotism and favouritism were perfectly natural, as was the assumption that only an officer could fly an aeroplane and only a gentleman could be an officer.

Grosvenor cannot be dismissed as a grandiose commander or a pleasure-seeking flyer. He realised that air fighting meant killing and, perhaps, being killed. As with all Territorials, patriotism placed at least second to *esprit de corps* as a motivation, and a commitment to fight for one's country was total. During the First World War pilots carried with them a revolver, as much as anything with which to shoot themselves if plunging to earth in a bonfire of wood and fabric and dope, but also as a means of attack. Exasperated by the ineffectiveness of this, Grosvenor carried a loaded, sawn-off shotgun in his cockpit; no ordinary shotgun of course, but a bespoke model made to his specifications by J. Purdey and Sons and designed for shooting at flocks of birds.

Although the navy and army looked down on the regular RAF, the latter never suffered from lack of self-esteem. Far from it. They knew that pilots were the cream of the military because only they could fly. Anyone can stand on a ship's deck or hold a rifle, they knew, and no doubt some sailors and soldiers had the potential to fly, but in reality they could not. Only pilots could fly, and there was no changing that. They could look down on the other services, literally from above. The Auxiliaries would adopt this self-regard and inflate it. Being from the cream of society they considered themselves double cream, as it were, and they were not bashful.

The Trenchard thesis called in the main for pilots who had no service experience. Aviation was still wondrously new, even suspect, and flying tuition was costly. The squadrons couldn't accept officers unless they had acquired a minimum standard of competence in the air already, so the Air Ministry offered to refund an applicant's tuition expense of £96 when he obtained an 'A' licence on his own account. This was the estimated average cost of the dual and solo training necessary before a pilot could be expected to handle a light aeroplane with reasonable safety. It was then up to the squadrons to elevate the new pilot to operational standard.

Each squadron had a regular service adjutant, and 601's first was Horace George, known as 'H. G.' Bowen, later Air Commodore. 'I know nothing about it,' apologised the senior officer who was supposed to brief him. 'You'll have to find your own way about.' Luckily, Bowen had known Grosvenor in France when flying with the Fifth Brigade, RFC, and respected Grosvenor. He reported at Northolt as Grosvenor's right-hand man. He had, among his many duties, to convert the new pilots to military machines, and would give them dual instruction in the Avro 504K before letting them loose in the replacement Airco 9As, flat-nosed biplanes with 400 horsepower Liberty engines, which then seemed a great deal of power.

At the first annual training camp at Lympne airport, held in August of 1926, only four pilots had any service experience besides Grosvenor and the two regulars. The location of Lympne was very agreeable. The surrounding Kent countryside was beautiful, with the downs rising to the north and the land sloping down to Romney Marshes and the English Channel to the south. There were two local landowners. Goldenhurst Farm was owned by Noel Coward, while Porte Lympne was the seaside residence of Sir Philip Sassoon, a friend of Grosvenor's. Both homes were permanently open to the officers.

Sassoon, a grandson of Baron Gustav de Rothschild of the Jewish Rothschild banking dynasty, a wartime second lieutenant and former member of the East Kent yeomanry, was serving as Under Secretary of State for Air. He quickly befriended the squadron, which couldn't have had a more influential ally, and entertained its officers lavishly, so that the annual camp came to be dubbed 'the Summer Outing of White's Club'. The term was more than a joke since waiters from White's would serve in the tented mess when the club was closed for the season. Sassoon wasn't only rich but commensurately generous. His homes were in Park Lane, Mayfair; Trent Park, Hertfordshire; Poona, India; and Porte Lympne. Trent Park was a palatial dwelling in north London, on extensive grounds that had in the fourteenth century been a hunting ground of King Henry IV.

But Porte Lympe, Sassoon's summer home, was his favourite. This was a huge and lavishly furnished historic mansion. The dining room walls were covered with rare stone, its ceiling with luminous opalescent stone, and the table was set with gilt-winged chairs and jade-green cushions, all surmounted by a frieze of scantily clad Africans. Statues of naked men in classical poses stood on plinths in the garden. While it would have been scandalous to suggest so then, the contemporary mind will have little trouble discerning the signs of a gay man.

The house, set in landscaped gardens designed by the architect Sir Herbert Baker, was a honey pot for the leading establishment and society figures of the 1920s and 1930s, including the Duke of York, England's future king; Charlie Chaplin; Winston Churchill while a backbench MP; Anthony Eden; Sir Samuel Hoare; sundry foreign ambassadors; and T. E. Lawrence ('of Arabia'), also known as 'Aircraftman Shaw', the celebrity with the lowest rank in the RAF by his own choosing. When a squadron member dined at Porte Lympne he would be among the most privileged of guests, for Sassoon's admiration for his pilots matched his regard for anyone else. These happenings couldn't escape the press, and it was inevitable that 601 should be tagged 'The Millionaires' Squadron', a name that stuck with it forever. The regular RAF would call it 'The Millionaires' Mob'.

With wide terraces set in sumptuous grounds dotted with life-size bronze statues of animals and classical figures, leading down to a swimming pool and tennis courts bordered by tall juniper trees, Porte Lympne was a splendiferous paradise close to Lympne airport, where the squadron camped most summers. Whatever one might say about Porte Lympne's Hollywood Baroque style – and behind his back many guests snickered – 601 loved it there. The officers and sometimes the NCOs and airmen came to swim, play tennis, lounge in deck chairs, dine, and rub shoulders with an eclectic assortment of VIPs from theatre, films, society, and politics. In the days of limited news, dining with Sassoon's guests and hearing it from the top was considered learning from listening, as some officers said.

One of Grosvenor's first tasks, into which he threw himself with diligence and pride, was to design a distinctive badge incorporating the idea of his squadron's link with the County of London. Being a tolerable artist, he drew on the back of a large official envelope a scarlet Sword of London sprouting wings at the hilt. It was a marvellous design of unusual economy: all red and representing flight, the County of London, and perhaps bloody battle. This came to be known as The Flying Sword. Due either to an oversight or a stroke of genius Grosvenor didn't add the customary ribbon under the crest bearing a motto. This design he sent to the Royal College of Heralds with a request that they arrange his device in a way formally acceptable to their standards. For a fee of five guineas the college made some minor alterations of an aesthetic character and entitled the finished product 'the Sword of London piercing a pilot's wings'. They raised no objection, and perhaps overlooked, the absence of a ribbon, an oversight they would come to regret. The 601 badge was and remained arguably forever the most elegant, captivating, and famous in the entire RAF.

Grosvenor set the pattern for the development of the squadron, and chose his officers from among gentlemen of sufficient presence not to be overawed by him and sufficient means not to be excluded from his favourite pastimes – eating, drinking and White's. He selected as potential officers those who would fit naturally into the setting of White's. He liked to see initially whether their social training had equipped them to deal with the large glasses of vintage port he would pour for them at his home, and if they responded satisfactorily he would take them to White's for even larger gins. He liked a couple of glasses of Marsala before breakfast to start the day in a civilised way and would frequently offer his officers a large brandy in the mess at Northolt. From his kindred spirits in White's he drew the first handful of officers before casting his net wider. It was said that he could have had more dukes and their heirs as pilots than the rest of the air force could muster collectively. But merely to have high social standing wasn't enough when it came to joining what Grosvenor was determined to make the finest volunteer unit in the country.

There seems to have been little of the democrat in Grosvenor, to judge by the earliest group photograph in which he sits pasha-like in front of an aeroplane hangar, central among officers, NCOs and airmen, haughty and humourless. But that would misjudge him. This photograph didn't project the casual, humorous, fun-loving personality that endeared him to every officer and airman once they had overcome their initial awe. Photographs were not commonplace and doubtless Grosvenor was investing this one with unusual solemnity, a formal record of a magnificent event, for he wished fervently for his squadron to be taken seriously. Other photographs show a more human face, a more relaxed individual. Whether by instinct or design, he created a squadron that always balanced its off-duty irresponsibility with great seriousness in this intent. He was deeply admired, even loved, by his followers.

The character of 601 took gradual shape as Grosvenor refused at first to delegate the task of selection. It became a curious 'soviet', as they would call it, its discipline no tighter than required by common purpose and respect.

Each officer and man who joined Grosvenor was impressed by his honest, magnetic personality and became his devoted friend.

He was altogether a strange man. It is hard to credit that, for all his love of flying, Grosvenor had an aversion to motor cars which prevented him ever driving one. He employed a faithful ex-member of his RFC squadron at the substantial sum of £2 a day to be on call, day and night, to drive him wherever he wanted in an upright Morris taxi.

Such was Grosvenor's charm that those who knew him would never tire of his classic stories about the French Foreign Legion. He had a great affection for the French, both individually and as a nation, and spoke quaint French and Italian. To him the adjutant of 601 was always 'Camp Commandant' or '*Commandatore' del Campo*'. He called night flying '*élévation grande distance de nuit*'. It was, in fact, while noting the continental-sounding names of the first clubmen to join him that he made a casual decision which stuck. 'This sounds like the Legion', he mused as he read Belleville, Driberg and Schreiber, then a moment later, 'Mon Dieu, this *is* the Legion, and we're all Legionnaires!' Not long after, a silver replica of a Foreign Legionnaire charging with a rifle in his left hand and a flying sword in his right would decorate the mess table.

The 'Legion' continued to attract young men incurably infected by the flying virus that spread in the late 1920s. They flocked in their own small aircraft, sometimes with struts literally attached by wire and the fabric held together by sticking plaster, to fly bombers at Northolt each weekend. Grosvenor's pride in his Legion was never greater than when he led its DH 9As in close formation. No pilot saw the back of his head for longer than a few seconds as his eye constantly roved over the other aircraft, and anybody deemed guilty of poor station keeping would be certain of an uncomfortable review on landing. Although very close to his officers in the mess, he never hesitated to make his orders clear, and they were backed by a formidable sanction: orderly officer duty. This, since it prevented flying, was a great indignity.

Brian Thynne, an early member, says: 'The job of orderly officer was very unpopular since you had to stay on the camp all day and yet were not allowed to fly. Moreover you had to remain in full uniform no matter how hot it was.' There were amusing consequences.

> The names for duty were on a roster beginning with the most junior officers, but days were often doled out as punishments, so the people at the end of the list got off to make room for others who had misbehaved.

> One year Drogo Montagu was found mounting the guard dressed in an open-neck cricket shirt and black-and-white golf shoes, and was promptly made orderly officer for a whole week to the delight of the seven officers who would have had to do it otherwise.

That the Legion in its first year won Lord Esher's trophy for the most efficient squadron can hardly be counted one of 'Ned' Grosvenor's great achievements. Once he got started, no other possibility ever crossed his mind.

If a conservative is one who doesn't want anything to happen for the first time, that couldn't apply to Grosvenor or his followers. While in the main socially conservative they welcomed the incredible new phenomenon of flight and relished every new design and technological advance that came their way. They were not satisfied to be proficient: they strove to be the best, and could be seen in all directions from the airport practising stalls, rolls, and loops. It wouldn't be too much to say they were flying fanatics. They flew their own planes to the airport and then they flew RAF planes. They flew whenever they could, as though this opportunity might suddenly be snatched from them. Flying bombers added purpose to pleasure. Close formation called for concentration and discipline, and they found this highly satisfying. At the1932 air pageant No. 601 displayed an amazing single close Vee formation of twenty-nine Wapitis.

Every Monday and Thursday evening the squadron would congregate in its substantial Town Headquarters at 54 Kensington Park Road, Notting Hill Gate, a large house with an imposing archway set in a tall brick wall, soon to be adorned with a metal Flying Sword. Inside were rooms for training on airframes, guns, and engines, and a drill hall. The officers' area contained an elegant green and gold anteroom with small framed military prints, a bar, kitchen, and dining room. The silver was exquisite by any standards. There was good food and wine at the long mahogany dining table when the candelabra had been lit.

Two games that developed naturally in the anteroom of Town Headquarters became traditional. The first was discovered when John Hawtrey, then the regular adjutant (and later air vice marshal), attempted to circumnavigate the room without touching the floor, a favourite RAF pastime. He found himself baulked by the substantial shelf protruding above the door, and only by leaving the door open was he able to cross the shelf. One night somebody shut the door while Hawtrey was on the shelf and left him stranded twelve feet from the floor, hunched up with his head touching the ceiling. On an impulse the other officers squirted soda water at him until he was drenched, and he was to see this develop into an initiation ceremony for new members, who would be artfully coaxed into mounting the shelf by a well-staged discussion on the impossibility of reaching such a high protrusion. For many years it was astonishing how easily the new man, eager to make an impression, could be manoeuvred into this position. No persuasion was ever used; the best occasions were those where a clever pretence at dissuasion 'failed'. Those who hesitated for any reason would fall for the approach to the commanding officer: 'Is it all right if John has a go at the shelf?' and he would reply thoughtfully, 'Yes, provided nobody helps him.'

'Calibrating the Table' needed a stooge and a victim, and there was never a shortage of either. The stooge would lie face upwards on the oval table in the middle of the anteroom, hanging by his heels as it was tilted back. There would be a show of measuring the angle and surprised comments: 'Thirty-four degrees – two more than the record!' When the victim tried it he would be tilted to an angle from which it was a simple matter for three or four strong young men to whisk him off the table by the ankles and hold him inverted while several tankards of beer were poured down his trouser legs.

Though 'Calibrating the Table' wasn't Hawtrey's invention, he was remembered with affection for his eccentricities. When he lost the key to the squadron safe he turned out all members to advance yard by yard across the aerodrome until it was found. He then lost the key again, and this time the unceremonious parade couldn't find it. Hawtrey had the safe loaded on to the back of his little Austin Seven and drove to London with his mudguards scraping the wheels in search of a locksmith. The Air Ministry sent him a bill for a new key and for the damage to the safe when it was forcefully opened.

Early recruits from White's included Richard G. ('Dickie') Shaw, unique in having a DFC from the First World War, Norman Jones, and the flamboyant Rupert Belleville. Grosvenor went about recruiting gentlemen for his squadron as for an exclusive gentlemen's club, which of course it was. Brian Thynne's path to membership may have been typical. His resume of Eton, Oxford, Officer Training Corps and Sussex Yeomanry plus family connections at the highest level were only starters.

Hating the noise and mechanical efficiency of the internal combustion engine and having no urge whatever to fly, Thynne, who much preferred horses, had never been in an aeroplane when he came down from Oxford and joined the Sussex Yeomanry. The son of a colonel in His Majesty's Bodyguard, his greatest asset was an unusually nimble brain. Ideas flowed so fast that Thynne's fluent powers of expression were taxed with delivering them at three hundred words a minute. He wore a thin moustache and was always immaculately turned out. The family printing firm for which he worked obtained a contract to print bank notes for the Chilean Treasury, which needed the notes to bear elaborately engraved portraits of national heroes. It was this fact that introduced him to flying. The only artist who could do the work lived in Paris, and to him time meant nothing. Thynne was forced to make weekly visits to the Paris studio to encourage and cajole the man into a compromise between art and expediency. He found the repeated land and sea journey tedious and 'not knowing then what Paris is for', as he put it, he turned to Imperial Airways for a means of shortening the travel time. He was delighted to find that he could do the return trip in a day and he didn't have to pack a suitcase. His only previous flying experience had been a joy ride for ten shillings, which he loathed. This, in an Argosy, was different and better, prompting him to take flying lessons at Shoreham. After four flights Thynne fell under the flying spell. He had had a dozen hours when the Sussex Yeomanry's horses were replaced by mechanical transport. That was enough.

Thynne learned about the Auxiliary Air Force and was intrigued enough to decide on a transfer from the yeomanry. His father took much persuading but eventually gave in. He resigned his commission and with his father's introduction applied to 'Ned' Grosvenor.

In selecting his officers Grosvenor used the standard approach of an elite cavalry regiment. He had no other guidance. The main question was, will he fit in? If so, all else would follow. If not, *esprit de corps* would suffer. Fitting in naturally meant being of the same social class as he. When Grosvenor interviewed Thynne at his home he didn't seem to care too much for what he saw because, Thynne believed, his suit was shiny and his shoes scuffed. 'I can't remember what happened', he recalled, 'except that I was extremely frightened of Lord Grosvenor and of Bill Thornton the squadron adjutant who was present.' Eventually Thynne's obvious enthusiasm and persistence in following up pulled him through, but then only on probation. Thornton later told Thynne, when they had become great friends, that they had decided to make it difficult for him to join, but that if he refused to take no for an answer they would let him in. Thynne would come to greatly admire Grosvenor's personality and leadership, even to love him as a man. He would also become the squadron's fifth commanding officer.

Thynne's first impression of service flying was that it seemed so safe. By comparison with what he had known the aircraft were sturdy and stood up to the heaviest landings. Training was thorough and step by step, and a student had to master aerobatics as well as take offs and landings before being allowed to go solo in order to be able to recover from any attitude, whereas civilian training permitted a solo after a few successful circuits. There was also a speaker tube enabling instructor and pupil to communicate without having to signal by hand. He was petrified of his instructor and brother officers, which he found a good thing because it took his mind of any fear of flying. Still, he was in a hurry, and he proceeded with characteristic cunning since 'my gazetting to the squadron would date from the time I got my "A" licence, and meanwhile I knew that Loel Guinness, who could already fly, was in process of joining, and I was determined to beat him to it and therefore be senior to him for ever and a day. In fact I beat him by a day, since he didn't know of my existence and wasn't hurrying'.

Inspired by his first solo, Thynne resolved to acquire his own aeroplane, and after much research bought a Simmonds Spartan. The market price was £620 (equivalent to £36,000 today) but he persuaded the dealer to accept £200 down and the rest in installments – perhaps the first aircraft sale on credit.

Thomas Loel Evelyn Bulkeley Guinness, an heir of the banking branch of the Guinness family, followed Brian Thynne. A competent flyer with his own private aircraft, though much to Thynne's joy, as he didn't want to be pre-empted, not yet up to military standard, he had been to the Sandhurst military college and served as a lieutenant in the Irish Guards. Guinness was to have a sensational private life played out in public. He would have, one can't say enjoyed, three celebrated marriages to famous socialites. His first and most famous was to Joan Yarde-Buller, daughter of a baron, whom he divorced eight years later after naming Prince Aly Khan as co-respondent. Mrs Guinness left him for the prince (who thereafter married the actress Rita Hayworth). Guinness won a seat in Parliament in 1931 and would later be named private secretary to the Under Secretary for Air when that appointment came for the second time to Sir Philip Sassoon.

Despite Thynne's impression, flying was far from safe by present standards. The aircraft were flimsy, engines unreliable, instrumentation negligible, crash-protection little understood, and there were neither protective helmets nor parachutes. (It would be thirty years before 'hard hats' were introduced.) The privately owned aircraft were even less safe. There were bound to be tragedies. Whitehead Read, a pilot from the First World War, bought an SE5 (First World War biplane fighter) and became one of the earliest private aircraft owners in the country. A surgeon living in

Canterbury, after joining 601 he would fly to Hendon at weekends in various unsafe-looking contraptions, arriving regularly every Sunday morning regardless of the weather. Over-confidence was to be his downfall when he killed himself flying a Wigeon into a tree in bad visibility.

The gulf between the squadron's two classes was clear. The officers and 'other ranks' differed sharply in background and lifestyle and had quite different civilian occupations. They were just as differentiated at squadron functions as in civilian life. A photograph taken in the mid 1930s of a squadron tea party after a sporting match, possibly cricket, shows the NCOs and airmen sitting at trestle tables on long wooden benches, as many as twenty in a row, in uniform, with boots, tunic buttons fastened, while officers in double-breasted civilian suits and ties lounge at the fringes. None of this caused resentment or guilt; it made for stability. Nobody expected things to be different; this was a model of society that seemed to work. The NCOs and ordinary airmen were enthusiastic volunteers, winners of a demanding selection process and deeply proud of their skills in airframes, engines, rigging, or gunnery and their membership of an elite unit. They lacked the education and means to learn to fly, and not everybody wanted to; but they could work with aeroplanes and develop knowledge and skills that were valuable in civilian life. Some would qualify to fly in the rear cockpit as combined bomb aimers, gunners, and observers. They were indispensable in a bomber, and aeroplanes were a unifier. A pilot and his ground crew were a close team and stayed together permanently until one of them died or left. Many airmen and NCOs would continue with 601 for years longer than any officer. There was even dormant pilot and officer potential among these ranks, which only the necessities of war would later uncover.

To signal the squadron's individualism the officers lined their uniform tunics and civilian jackets with bright scarlet silk and wore blue ties in place of the regulation black. They played polo on brand-new motor cycles and drove fast sports cars, the parking area at Lympne resembling a *concours d'elegance*. Although they often mocked the regular RAF, that was a pose. They flew RAF machines and were part of the RAF's combat readiness. Each squadron also had regular adjutant, plus a deputy adjutant (later training officer), on both of whom they leaned heavily.

The behaviour of the squadron never ceased to be undergraduate in its off-duty expression, and of this Grosvenor thoroughly approved. He had an uncommon sympathy with the adventurous and the eccentric, and perhaps nothing can illustrate his attitude to men better than the case of Corporal T. Wewege Smith. The high quality of 601's non-commissioned ranks was legendary in the air force, but Corporal 'Wiggy' Smith was even more an individualist than his comrades. He arrived in a Daimler at the age of nineteen to join the Legion as a trainee air gunner. After an unhappy love affair and inspired by the example of his commanding officer Smith left England abruptly to join the French Foreign Legion with the mental reservation that he would escape if he didn't like the life. He wore his best plus fours when presenting himself at the Dunkirk recruiting office and was a little surprised when the officer in charge suggested he should have a good lunch and consider carefully what he was about to do.

After six weeks' recruit training and road construction in Saida, Smith realised how good that advice had been. He decided to escape and said so to the English medical orderly, who suggested that Smith's eyes were none too good (they were perfect). He never discovered how many wires were pulled but was eventually discharged as unfit for further service. Only then did he begin to realise that, technically at least, he was liable to severe penalties for deserting 601. The adjutant, then Flight Lieutenant (later AVM) Thornton, wasn't in the least pleased to see Smith when he reported back for duty at Hendon, the squadron's new airfield. Thornton had reached the climax of his outline of the penalties that might befall Smith when Grosvenor entered the orderly room. 'Good God, Smith!' he roared delightedly. 'So you've been out and come back to us. Marvellous show, and, by Gad, you're looking damn' fit.' After an exchange of yarns about Saida and Sidi Bel Abbès, Smith's case was disposed of.

To the extraordinary Corporal Wewege Smith went the honour of being the first man to carry the Flying Sword into battle. For his wanderlust days were not yet over. After another love affair failed he went to Latin America and joined the Bolivian air force as an air gunner in the war against Paraguay. He found it helpful to have been a member of the squadron that won the Esher Trophy because his Bolivian comrades believed he had helped to win the Schneider Trophy (international seaplane race), and he didn't disabuse them. They awarded him the Order of the Condor of the Andes and paid him well as a mercenary officer bomb aimer. Smith painted the Flying Sword on the nose of the Junkers Ju 52 in which he flew his first raids. In 1936 Smith returned to England. His arrival at Hendon was dramatic. One Sunday morning he emerged from a taxi, a gorgeous figure with shining eagle-crowned helmet, a cape of two colours slung carelessly from his shoulders, a short sword and tall boots, outdoing the Life Guards.

It was unfortunate that 'Ned' Grosvenor was unable to enjoy the spectacle of Corporal Smith's second return from his adventures, for he died suddenly at the age of thirty-six after a short illness, in 1929. It was a deep and unexpected loss for the Legion. The need was felt for a permanent reminder in Town Headquarters and his portrait was hung over the fireplace from which position it smiled – cap at an angle, in baggy riding breeches and bemedalled tunic, standing in front of a Wapiti bearing the Flying Sword – upon mess meetings, guest nights and the ceremony of the shelf, for many years to come.

Somewhere, nobody can recall where, the phrase 'Spirit of Grosvenor' came into being and stuck. It has never been defined. It might have meant something like, 'Be a good pilot and a good comrade, but don't take anything other than flying too seriously.' Thynne believed it meant presenting a first-class example of British qualities wherever in the world your travels might take you. Whatever the phrase meant, the Legionnaires believed they were living up to it. The instantly recognisable Flying Sword icon continued to be a focus of loyalty for Legionnaires of all ranks and multiple nationalities, on aircraft tails and noses, cars and lorries, lapel badges, ties, buttons, and cuff links, even as ruby-studded silver brooches for wives – or girlfriends who had 'earned their wings' as it was put, and as a hands-off signal – for many years to come.

Chapter 2

Fun and Games

> I landed and taxied up, feeling as all first soloists do, that if I died the next day at least I had achieved something in my life.
>
> Brian Thynne

The squadron was lucky when on Grosvenor's untimely death Sassoon, having left office as Under Secretary for Air on a change in government, replaced him as its second CO. He was well accepted as Grosvenor's successor, bringing to it a bridge to influence with the Air Council, a political shield for some of its wildest mischief, and not least a limitless generosity. But he was unique as a squadron commander because, despite his stable of privately owned aircraft and the airstrip he had built in the grounds of Trent Park, for a long time he simply couldn't fly. It wasn't for want of trying; he simply had little aptitude in coordinating the several physical actions involved while thinking clearly of his relationship to the ground. In this he wasn't unlike any other student pilot but he showed little sign of making progress. He couldn't even drive a car very well. In the end, he had to be 'helped' to earn his pilot's wings. Sassoon brought a touch of comic-opera to the Legion with his lavish generosity, his eccentricities and his dogged attempts to fly.

His quaint expressions were often memorable. A guest, Sir Henry 'Chips' Channon, recorded one in *The Diaries of Sir Henry Channon*: 'It's so quiet at Lympne you can hear the dogs bark in Beauvais.' The flowers in the grounds of Porte Lympne were planned to emerge in splendid profusion at the one time of year when 601 encamped at Lympne, and a lady visitor once asked how such synchronisation was achieved. 'Oh, it's perfectly easy,' explained Sassoon, with an elegant wave of the hand, 'at twelve o'clock on the first of August each year I give a nod to my gardener. He rings a bell, and all the flowers pop up!'

His charm, his soothing manner and sibilant, sing-song voice were captivating. It was to the other ranks that Sassoon projected a less successful image, perhaps because of his darkish complexion, being of Middle Eastern,

Parsee descent, or his oriental love of display, or a suspected feminism. Or maybe he didn't really try.

Sassoon placed his aircraft, ranging from a Leopard Moth to a Percival Q6, and the services of his pilots upon whom he leaned heavily in such matters, freely at the disposal of his many friends. On occasion he would telephone a pilot and ask him to fly Anthony Eden to Paris, or someone to the South of France. When a young pilot agreed to fly Miss Amy Johnson to the Continent Sassoon, for some unknown reason, gave him firm instructions that on no account was the world's most famous aviatrix to be allowed to touch the controls.

Despite his wealth Sassoon seemed to feel that the squadron contributed more to his happiness than he did to theirs. He was pleased, proud even, to be accepted as the leader of a group of Auxiliaries. Grosvenor had been the raconteur, the founder of the Legion, the pilot with the Blériot and Purdey gun. Sassoon was mild and sensitive, the successful politician, the unsuccessful pilot. But despite his generosity, even his saintly patience must have been tried occasionally. In one squadron game a hip bath was towed by motor car at high speed round Lympne airport while its occupant, clothed in protective Sidcot flying suit and leather trousers, remained in the bath as long as possible. The hip bath would swing widely at each turn and brush its passenger against the ground. One evening the squadron doctor, who usually stood around in case of mishap, attempted a circuit himself but was pitched out of the bath head first on the concrete letter 'L' forming part of the sign LYMPNE. The value of head protection wasn't then or for many years recognised, and he was knocked unconscious.

Sassoon took no part in these activities but redoubled his efforts to fly. Almost in despair, he asked the Chief of Air Staff (CAS), Sir John Salmond, to dinner and then enquired whom he thought was the best instructor at the Central Flying School, where instructors taught other pilots to be instructors. 'There's a promising fellow named Boyle,' replied Salmond after some thought, 'a first class pilot and an officer with a future'.

Two weeks later Flight Lieutenant Dermot Boyle was posted to 601 as regular adjutant. Somebody asked him, shortly afterwards, how he was getting along in teaching Sassoon to fly. 'I'm not teaching him to fly,' Boyle corrected, 'only to land.' Nevertheless, Sassoon finally gained his wings under Boyle's expert guidance, going solo after more than twice the usual number of hours' instruction. He made his first solo at Cranwell, which, it was noted, had the longest runway in the country.

Boyle got on well with Sassoon from the start. The start happened to be an instructional flight during which Sassoon quite unexpectedly interrupted the training sequence by ordering Boyle through the speaking tube to fly over Blenheim Palace. Boyle was horrified. 'I can't do that sir, it's a prohibited area.'

'Oh, rubbish!' replied Sassoon sternly, 'Blenheim, I said. I have a friend who's thinking of buying the place.'

Since his pupil wasn't only his CO but about as significant a political figure as could be, Boyle felt reasonably well covered, mentally shrugged, and complied. They circled Blenheim Palace at a demure two thousand feet.

'Lower, Boyl-o!' shouted Sassoon, 'lower!' Then, 'Very nice little place, eh, Boyl-o!' and, as they climbed away, 'but did you notice that the kitchen garden looked a little pinched?'

For many regular officers an appointment to No. 601 Squadron as adjutant or training officer proved over the years to be a ticket to high rank. 'Dab' Boyle was one of these. With his great sense of timing, he was one of the six best pilots in the RAF, destined to become CAS. When he put on a show he impressed all members of the squadron. Having started up his Lynx Avro (an Avro 504 with a Lynx-Avro engine) he would let the tail rise until his propeller was cutting the grass during the take-off roll. His manoeuvres were well placed down sun, often so low that his wingtips collected gorse and mustard from the hedgerows. Sometimes he would attach wooden sticks to his wings and break them off with two vertical turns a few feet above the ground.

Sassoon's training at Cranwell wasn't concluded without incident. George ('Geordie') Ward, a young Auxiliary, received a phone call at Hendon to say that he had flown through a hedge on landing and would be obliged to receive a replacement aircraft. Ward flew up another Avro. He found Sassoon standing beside a crumpled wreck and asked if he was all right. 'Thank you, Geordie, I'm OK', replied Sassoon in his inimitable accent, 'but I'm afraid the aircraft is confetti!'

Sassoon decided to present the squadron with a private aircraft, and Thynne was appointed to make the choice. He bought a Spartan, a three-seater open biplane, which cost Sassoon about a £1,000. The flying costs per hour were calculated and the pilots would book it, much as they would a squash court, by writing their names on the notice board at Hendon. The insurance on the Spartan was considered rather heavy, and since there was no claim during the first year but the insurers were only prepared to drop the premium from £75 to £70, 601 decided not to reinsure.

They couldn't have considered probabilities. Thynne and John Gillan (who at two o'clock one morning decided on an impulse to fly to Dubrovnik in Jugoslavia and did so with a school atlas) flew to dinner at Thynne's parents' home in Sussex. Thynne took the Spartan and stayed overnight. Gillan flew in Sassoon's Moth, which he needed to return to Lympne that night. By the time Gillan arrived at Lympne it was dark, and he landed on top of one of the numerous cows belonging to a local farmer that were allowed to graze on the airfield overnight. On his arrival in the Spartan next morning Thynne was so amused to see the Moth's bizarre predicament that he landed as close to it as he could. The Spartan lacked wheel brakes and trickled into the rear of the Moth, splintering the propeller. When Sassoon protested at the damage to his Moth, Thynne replied with feigned indignation that it was the Moth that had reversed into the Spartan. Sassoon, delighted with this explanation, paid both bills.

Something far less amusing was to happen with the Spartan before the year was out. Simon Gilliat booked it to bring his brother, Lieutenant John Gilliat, from the army camp at Aldershot to a squadron guest night at Lympne. He landed at Farnborough where his brother was waiting, but made an error many pilots have made. He parked the aeroplane with its nose into wind at the duty hut, but while he was away the wind veered completely round. On take-off downwind the Spartan failed to clear some telegraph wires and crashed on to the airmen's cookhouse, bursting into flames. Simon Gilliat escaped from the machine but was prevented by the heat from rescuing his brother, who died in the flames.

When Thynne learned the extent of the damage, including a claim by the Air Ministry for £3,000, he promptly telephoned the insurance company to explain that the cancellation of cover hadn't related to third party. The insurance company didn't quibble and promptly settled the claim.

On a training flight Thynne was horrified when his gunner, who had been crouched frozen at the bottom of his cockpit and in rising accidentally pressed the Salvo Release button, uttered the words, "Oh my Lord, they've gone!' The practice bombs fell into a close pattern between the ornamental lake and the drawing room windows of Blenheim Palace. Flying over the palace was forbidden, 'let alone bombing it', as Thynne said. However, the occupants must have been away because nothing more was heard about it.

Sassoon liked large formations. He arranged for all the private aeroplanes at Lympne to take off and fly in formation to a field at the stately home of a friend for lunch. It was supposed to be a display of precision flying. The assortment of nine machines, led by Sassoon in his Puss Moth, carried among its passengers Aircraftman Shaw (as T. E. Lawrence had bizarrely reinvented himself), and Winston Churchill. Sassoon loved to fly at low level, which Churchill steadfastly refused to do, with the result that the formation when it arrived had a somewhat ragged shape that was less impressive than Sassoon had hoped.

With Grosvenor gone, applicants were now required to have a one-on-one vetting dinner with Sassoon. Christopher John Henry ('Jack') Riddle and his older brother Hugh Joseph (contracted to 'Huseph') Riddle, a talented artist, went through such an interview and, like Thynne after his grilling by Grosvenor, were too paralysed with apprehension to remember any of the conversation afterwards, though they obviously passed muster. Given subsequent events it is hard to think of Thynne and the Riddle brothers being afraid of anything, yet the prospect of being turned down by 601 Squadron seemed to them a frightening one. High birth was less a requirement than it had been, though it certainly wasn't an impediment. A public school education followed by Oxford or Cambridge and proficiency in athletics or a sport such as skiing, rugby, or cricket, were desirable criteria. As for flying ability, that's what the regular deputy adjutant, later training officer, was for.

Sassoon's flying gradually improved, and the Legion's competence as a bomber squadron progressed until, it was said, almost every pilot was able to drop almost all his bombs within the danger area on the range. War was still a remote possibility and little attention was paid to sophisticated bombing techniques. Whenever the weather was suitable for bombing practice the ranges were usually fully booked, and so a synthetic trainer, the Bygrave Teacher, was supplied to all bomber squadrons. Using a motion projector, the Bygrave simulated a moving countryside and produced artificially all the problems of altitude and drift that a pilot might encounter on a real attack. The Legionnaires knocked a hole through one of the floors at Town Headquarters and installed the projector to produce a landscape on the floor of the room below. They built a dummy cockpit with pilot's seat and rudder bar, and fitted a plumb bob that moved across the landscape according to the estimated velocities of aircraft and wind. If the pilot pressed his release button at the correct moment the plumb bob would drop on to the predetermined target. There was always a sweepstake to be won by the pilot with the best evening's performance, and since there were sixteen pilots and a good number of side bets it wasn't long before the casino spirit obliterated the military and it was often well after midnight before practice ended.

The Avro 504 was a reliable and stable aircraft, generally idiot-proof for normal circuits and landings, and almost capable of flying itself. It was said to be so safe it could only barely kill you.This was, of course, a challenge in itself to the hyperactive Legionnaires. A contest began with a spotted handkerchief tied to the stick in the front cockpit of an Avro 504 dual-control trainer, which the contestant, flying from the rear cockpit, would somehow land with in his pocket after a circuit of the field, timing being from chocks away to touchdown. Because the handkerchief was sometimes tied so tightly it could scarcely be loosened on the ground, or the winner was too often the pilot with the longest arms, they found a way of dispensing with the handkerchief altogether. Now the pilot merely took off in the rear seat and landed in the front.

To avoid a poor time the pilot had to make his turns as quickly and as low as he dared, leaving the barest margin in time and altitude to avoid stalling. The refinement of this sport wasn't complete until two pilots actually took off in the same plane and changed cockpits in the air – a development so unexpected that it had to be repeated to be believed. This version of the game came to an abrupt end after an experienced pilot tried it with an *ab initio* pupil and failed to notice until after changing cockpits that there was no control column in the other one, so that the pupil, not wanting to risk another cockpit change, had his first landing as pilot-in-charge thrust upon him. The experienced pilot's reaction is gleefully recorded verbatim, but without attribution, in the Line Book: 'Christ! I've got no stick! For God's sake don't crash! For Christ's sake don't pile me up!'

For the first year or two after their creation, as the Auxiliary units felt their way, there was little time for feuding, and mischief was directed against their regular counterparts rather than against each other. When encamped at Lympne, 601 traditionally displayed a prominent notice, purporting to be official, outside the dispersal: 'AAF ONLY. TRADESMEN, RAF, ETC., ENTRANCE AT REAR. By Order …'

But it was inevitable that a rivalry should develop between the two London Auxiliary squadrons, 601 (County) and 600 (City). The seeds were sown on 14 October 1925 when both were simultaneously gazetted – along with three others, one in Birmingham and two in Scotland, which were too far away to worry about. The question arose: Which squadron was the first? No. 600 Squadron claimed that it had an adjutant before 601, and two of the other three squadrons were quick to chime in that they had adjutants appointed at the same time as 601. Number 601 declared that it was the first to have a commanding officer, so there, and such was its creed forever after. (In 1952 the Air Ministry's answer pleased neither squadron. It was that the appointment of an adjutant formed the first squadron, which made

No. 602 (City of Glasgow) the oldest. 'Some people are asking', grumbled the Legionnaire Max Aitken's *Evening Standard*, 'when is a squadron not a squadron? The answer might be, when it has no CO.')

So when Squadron Leader Peter Stewart, CO of No. 600 Squadron, generously presented a magnificent silver challenge cup to be competed for by all five Auxiliary squadrons in a relay race, the rivalry assumed a high intensity. The rules laid down that the machines should be standard Avros and that there should be no low flying, but according to Thynne 'nobody paid any attention to either rule'. The wings' angles of attack were flattened to induce speeds the makers had never intended, while every tall wireless aerial around managed to get knocked down. 'Never', reported *The Aeroplane*, 'have Standard Avros been so un-standard.'

After elimination races in the north the semi-final came down to 601 and 600, the winner of which would race the northern contender in the final. The course was a triangular one, beginning and ending at Hendon and punctuated by two turning-point markers consisting of large white crosses on the ground, at which observers were to be posted. When RAF officials visited Hendon to watch practices, some deliberate and rehearsed near misses were staged by 601, with Avros crossing and re-crossing each other within feet. This looked so dangerous that a ban was imposed on practices over the course, which was 601's crafty intent. There was another rule in line with 601's scheming, that if a pilot missed one of the turning points his team was disqualified.

To expect British fair play in the developing rivalry between 601 and 600 would be to expect snow in the Sahara. The Legion knew that despite the ban 600 would practise over the course, and they did, for they were determined to win the cup their own CO was presenting. The 601 pilots didn't bother to practise clandestinely – they knew the area anyway – but contented themselves between 600's practice flights and the race by moving one turning-point marker to another field, replacing it on the day of the race. No. 600 assumed they had won the race by miles and both squadrons were drinking their champagne – it took eight bottles to fill the cup – when a turning-point observer, Bill Langdon, 601's store officer, who had been delayed because his car broke down, arrived during the celebration to announce that 600 were disqualified because they hadn't turned over his marker. The commotion this evoked was understandably immense.

The final was between 601 and the northern semi-finalist squadron. Since 601 knew every inch of the course while the other squadron wasn't permitted to practise and had no time to anyway, the Legion won easily. The

Air Ministry was so horrified by the whole matter that future races were cancelled, which was fine with 601 because that meant the cup became part of its permanent silver. It wasn't even called the Peter Stewart Cup but the Langdon Cup, because of Langdon's timely disqualification of 600.

Confidently, 601 embarked upon the 1930 Air Defence of Great Britain exercise, when Auxiliaries joined regulars and the country was divided into Redland, the attacking bombers, and Blueland, the defending fighters. The 601 DH 9As had just been replaced with Westland Wapitis, somewhat ungainly biplanes, which were more advanced in many ways but not significantly faster and with less personality. It had a radial engine, a forward-firing machine gun, plus a Lewis gun for the observer, and could carry 580lb of bombs. There were umpires flying as observers with both sides and positioned on their bases, calculating any hypothetical damage and loss of aircraft. The Legion, from Redland, was ordered to bomb Yorkshire targets in the north-east corner of Blueland, and this they did well enough. Both on the way to the target and on the return, however, they were continually marauded by enemy fighters, Bristol Bulldogs of No. 54 regular squadron based at Hornchurch under the command of Squadron Leader 'Poppy' Pope. The pugnacious-looking, radial-engined Bulldogs were suitably named. The Legion's aircraft carried 'umpires' who decided that the bombers had been intercepted and destroyed before they even reached their target. These umpires annoyed 601, who constantly wrangled over the hypothetical losses.

This situation built up to an inter-squadron feud that lasted two years, ignited by No. 54 having followed the Wapitis to identify their base, then raiding it at first light each morning. The Lympne-base umpires decided that three or four 601 aircraft had been theoretically destroyed by the fighters, the penalty for which was that the first three or four 601 pilots scheduled to fly would be grounded for the day. This didn't sit well with individuals who had risen in the small hours in order to take off early. There was nothing that could be done about the umpires, who refused to allow 601 to strafe Hornchurch in return, but No. 54 itself was fair game. It was Sir Nigel Norman's idea to fly over in a formation of civil aircraft, over which the Air Council had no control, and bomb No. 54 Squadron with red ochre, soot, and anything else. Thynne remembered that rolls of toilet paper dropped from a height were said to unroll and cling to anything they touched. One evening, tests showed this to be the case, especially if a few feet were first unrolled to catch the air and pull out the remainder, and that of the brands tested Bronco was the best for this purpose.

Norman set off for Folkestone to buy a large order of Bronco, while Nigel Seeley put in a minority order for boxes of Jeyes toilet paper, cut into sheets, which he believed would take longer to clean up. The squadron had long been proud of its privately owned fleet of light aircraft of mixed types and makes while there were only about a hundred in the country, and that it could put a private force of mixed aircraft into the air, so this was a God-sent opportunity. Next morning the Legionnaires were taken by surprise when the exercise was abruptly terminated, with an armistice to take effect that day at one o'clock. In haste the motley formation took off at lunch time from Lympne. Thynne had wisely persuaded Sassoon to accompany them. The bombers arrived over Hornchurch just after the armistice went into effect, circled low to attract attention, then bombarded the spectators with hundreds of rolls of toilet paper – an estimated twelve-and-a-half miles – which unfurled in descent. The attackers criss-crossed the airfield in threes and their propeller wash swirled the rolls around until they stuck to everything: hangars, trees, telegraph poles. There was nothing the spectators could do about it, and the Legionnaires 'nearly died laughing', as Thynne recorded. As a final gesture Loel Guinness landed his Bluebird and placed an enamel chamber pot inscribed with well chosen messages addressed to 'Poppy' Pope next to the landing 'T', taking off again before anyone could reach him.

The Aeroplane ended its formal report on the air exercises with an account by 'an anonymous correspondent', which had for his friends the unmistakable style of Nigel Norman:

DASTARDLY VIOLATION OF THE ARMISTICE

The following is alleged by a correspondent to have been extracted from The Blueland Mail on August 15th.

An event of some significance is reported from Hornchurch Aerodrome, Blue Colony, of utter violation of the armistice declared at 1300 hours today, the hangars and buildings were subjected to a dastardly bombing raid of a barbarous nature.

The officers of the station had barely closed their eyes after the midday repast when a formation of nine aircraft was reported approaching from the S.E. It is said to have been composed of Moth fighters, Spartan day bombers, Bluebird cooperation machines and Pussaloons, and to have been led by a Widgeon interceptor.

After a swift reconnaissance of their objectives the enemy broke up into flight formations and carried out an elaborate plan of low bombing

> attack. A feature of this was the use of a hitherto unknown form of missile consisting of a central core or tune round which was wrapped many coils of a white material closely resembling the common bank note. When released these bombs uncoiled themselves in midair, descending upon the targets with widespread and disastrous effect.
>
> When eleven flight attacks had been made one of the raiders profited by the momentary disorganisation of the defence to descend and with unspeakable audacity to deposit a vase or vessel, resembling a domestic utensil, in the middle of the aerodrome circle. The object, which is now in the hands of the Intelligence Department, has been found to be inscribed in letters of the appropriate colour, with an ironical greeting from 'The Redland Forces'.
>
> No explanation can yet be given of this outrageous attack. It is suspected that the raiding aircraft were manned by ex-officers of the Royal Air Force. Rumour has it that the gang wasn't entirely composed of junior ranks.
>
> A sinister aspect of the affair is that although some 71,000 feet of the paperlike material has been recovered and forwarded to the Cypher Department at General Headquarters, so far no clue as be found to the origin and method of its manufacture beyond the well-known watermark which appears on every sheet.

This was rubbing it in. But 'Poppy' Pope had a good memory, and he waited a year for his revenge. No. 601 was on summer camp at Lympne when an illuminated scroll dropped from the sky, which announced formally that the CO of No. 54 Squadron presented his respects to No. 601 Squadron, and wished it to be known that his unit would be paying a call of an aggressive nature at 1200 hours the coming Tuesday. The message added, with barbed thoughtfulness, that it might be wise to clear the circuit of pupil pilots.

At thirty seconds to noon there was no sign of raiders at Lympne, which prompted remarks about regular pilots not making their ETA. Then, at 1200 hours exactly twenty-four Bulldogs in faultless line astern pulled up over the Channel cliffs and tore over the field at low level. On passing the signal square, each Bulldog jettisoned a tin chamber pot individually addressed to a Legionnaire. Within a minute the sky was completely clear again, and the 601 pilots cheerfully lined up to be photographed wearing the pots as helmets.

Of course, matters couldn't be allowed to end here. Some days later, while searching for on a form of retaliation, a Legionnaire found some fingernail-sized crabs on Folkestone beach. These were gathered in hundreds and packed in matchboxes, six at a time. Coloured streamers were attached and the boxes dropped, with an insulting explanatory note, on Hornchurch.

Sassoon was sensible enough to realise that the squadron's private aircraft relieved the Air Ministry of a great deal of responsibility. If the pilots hadn't used their own machines, decrepit as some of these were, they might have been tempted to use service aircraft and provide the basis of a few awkward supplementaries at Question Time in the House.

When The Hon. Drogo Sturges Montagu, the son of Lord Sandwich and one of the Legion's wilder figures, made a friendly dive over Trent Park in his Wapiti, forgetting to wind in his wireless aerial, so that the lead weight and twenty yards of steel cable crashed into Sassoon's ornamental lake, Sassoon was less concerned about the damage to his exquisite property than at the thought of what might have happened to the aircraft. Montagu, as it happened, excelled in impromptu air displays and was even audacious enough to beat up White's in broad daylight. He did so just as the CAS, Sir John Salmond, was emerging from the nearby Ritz with an Air Officer Commanding Group (AOC), with whom, over lunch, he had discussed flying and air discipline.

Wildness took many forms. Lord Knebworth, the eldest son of the Earl of Lytton and educated at Eton and Oxford, was elected to Parliament in 1931 while in his late twenties. That same year he joined 601, and on getting his wings the following year purchased a second-hand Moth that was in very poor condition. Members of the squadron who flew it said that 'some of the cylinders fire some of the time'. But Knebworth's enthusiasm for flying knew no bounds. He flew his Moth almost all the time and almost everywhere. He was certainly a refreshing character, and his flying was refreshing too. He could never see the point in limiting landing spots to aerodromes and would land in tiny fields, which shook his colleagues, considering how little flying experience he had. What he called his home aerodrome invited trouble, and when Thynne saw it – and refused to land there – he was able to understand how to Knebworth any other field looked easy by comparison: it was a small basin-shaped field surrounded by trees. Knebworth would take off upwind and downhill, hoping to clear the trees opposite as the ground rose up towards him. If that didn't seem likely he would do a tight turn, so close to the ground that he nearly dug his wingtip in, fly downwind, and hope to clear the trees at the other end. If that didn't seem likely either he would turn upwind again and, having by now inched up in height, finally clear the original trees. It was said of him that he had all the qualities of a first-class pilot except judgment.

Knebworth's most memorable landing was on the lawn in front of the drawing room windows at Cliveden, the Astor estate in Buckinghamshire and meeting place of the notorious Cliveden Set. Once there, even he realised he wouldn't be able to fly the plane out, and nobody from the squadron would risk landing there to offer advice. At a loss because he was a guest of Lord Astor that night and his Moth was all you could see from the window, Knebworth had the wings folded and his plane towed away.

Engine failure wasn't that unusual. Neither was it always that dangerous, since the slow planes could easily be put down in most small fields. But when a pilot force-landed a Wapiti on Hampstead Heath in bad weather, claiming his engine had 'faded', Joe Fogarty, the adjutant, had doubts, the more so when he went look at the machine and it started perfectly. Rather than have it dismantled and taken back to Hendon in pieces to be reassembled, he decided to fly it back. He selected the longest run into wind he could across the heath, but when about twenty feet up and too low for him to turn back the engine failed and the Wapiti crashed on the roof of a house, spread-eagled on the chimney stack with its wings hanging down drunkenly. The owner of the house was a doctor who took the whole matter with equanimity. Fogarty helped put out the kitchen and drawing room fires from petrol trickling down the chimney and the doctor was made an honorary member of the officers' mess, with whom he exchanged Christmas cards thereafter every year. According to Thynne, 'The only comment the authorities made was to tick Joe off for not wearing a parachute at the time, though what bearing this would have had on the accident is hard to see.'

Chapter 3

Send the Children Out of the Room Now!

That's the trouble with these Auxiliary squadrons. It makes us attached regulars forget all about discipline.

Flying Officer Elsmie, Upper Heyford
(Assistant Adjutant, 1932)

Upon another change in government, in 1931 Sassoon was restored to his earlier position as Under Secretary of State for Air and had no choice but to relinquish command of his squadron, but he became its Honorary Air Commodore. Sir Nigel Norman took Sassoon's place. Tall, athletic, with sharp, faintly eagle-like features, Norman was the son of a famous explorer, inventor and man of letters. Among the first of Grosvenor's intake from White's, he had learned to fly at Stag Lane, Edgware Road, in 1926. When he was proficient his mother gave him an Avro biplane, to which he fitted Handley Page automatic anti-spinning slots, being the first pilot to do so. Though he had a somewhat more sensitive nature than some of his colleagues – he wrote poetry for relaxation – he was by no means out of place among them, and shared and even led their boisterous antics. He had a far-seeing view of flying, which was rare, and anticipated the tremendous expansion of private and civil flying in the next twenty or so years. A member of the family gave him £50,000, enough in those days for retirement in ease, but instead he invested it in the development of Heston aerodrome, seeking to make the aerodrome buildings, living quarters and airfield equipment the most advanced in the country. His commitment was such that the Duke of York (later King George VI) formally opened the airport and his brother, then Prince of Wales, regularly flew from Heston. The squadron patronised Heston regularly with their private aircraft and were always accorded unfair priority at the petrol pumps.

On one of his visits to Heston the Prince of Wales forgot to bring his flying helmet and bought one from the aerodrome shop. On landing, he took off the helmet and noticed that it left a stain of blue dye on his forehead. Feigning indignation and holding out the helmet he protested, 'Norman, you sold me this helmet; now what do you intend to do about it?' 'I'm extremely sorry, sir,' Sir Nigel replied, 'but I think we can sell you something from the shop which will remove the mark.'

The Wapitis were replaced in 1933 by graceful and aerodynamic Hawker Harts, biplanes with shiny metal engine cowlings, sharply pointed noses, swept-back top wings set well forward of the lower, and large, sturdy wheels. Not only were the new aircraft gorgeous to look at and exciting to fly, but their metal framing represented a massive technological advance. Sassoon determined to have a Christmas card showing himself leading his famous unit in one of these beautiful machines, and was quite unembarrassed to have another pilot fly him from the rear seat of one with dual controls and then duck out of sight for the camera when the formation was in place.

There were grim reminders that, even in peacetime, flying warplanes wasn't always safe. Flying never can be completely safe, the aircraft were frail by later standards, and flight instruments were primitive. Bill Collett, one of Grosvenor's two original flight commanders and now CO of No. 92 Squadron, died in an air display.

In a worse instance, the circumstances were difficult to absorb. In 1933 Tony Knebworth, the wild and fearless one, died in a catastrophic accident that shook the squadron. Knebworth was in good form, and the tragedy had nothing to do with the aircraft. Nor was it Knebworth's fault, but that of the regular adjutant, Flight Lieutenant Eric ('Hobby') Hobson, a quiet and conscientious man who had replaced Dermot Boyle and could have hoped for a similar pathway to high rank.

Hobson was leading a close formation of Hawker Harts in practice for an air display at Hendon. Normally Nigel Norman as CO would have led the formation, but for business reasons he couldn't. Diving towards the airfield in one manoeuvre, Hobson inexplicably steepened his dive, held it too long, and then pulled up sharply. The manoeuvre was so abrupt – there was no radio communication, and no time for hand signals – that the other pilots, who had their eyes glued on him, were a fraction of a second late in reacting. It is possible Hobson was confused by a light mist hanging above the ground. All the Harts came very close to the ground and at least three struck it, including Hobson's and Drogo Montagu's, leaving deep ruts in the turf, but Knebworth's Hart hit it at a shallow angle; its undercarriage broke off, it burst into flames, and 'skidded on, almost keeping its place in formation', according to Thynne, leading another section.

'Hobby' Hobson was known to be a pilot of exceptional skill who exercised appropriate caution, and nobody could understand what happened. His potential for a successful career with the RAF was now shattered, but it is doubtful that he gave that a thought because he was consumed with guilt, like a man deranged, and had to be escorted home by two officers to explain matters to his wife. For days he lamented that he had murdered a man and couldn't be consoled. The court of enquiry found error of judgment on Hobson's part and, being a regular, he was posted way from the squadron. Some years later Flight Lieutenant Eric Hobson, while in the Middle East, shot himself.

That evening the Earl of Lytton drove out to meet with Thynne. It wasn't a meeting Thynne was looking forward to. However, the Earl was calm and courteous, and to Thynne's relief didn't ask to see the body. The air display went on as planned, and senior RAF officers graciously elected to occupy the gunners' rear cockpits in order to demonstrate their confidence in the Auxiliary pilots. The press reported Lord Knebworth's death in solemn detail but few gave thought to the gunner. One newspaper did, with grating condescension: 'As to Ralph Harrison we must also express our very deep sympathy with his relatives. Although only one of the minor men of the Royal Air Force he did his job just as effectively as any of the officers.'

After three years, during which he saw the squadron persevere in its training on the two-seater Hart biplane, Norman left to attend to his business interests, including the foundation of Airwork Limited. He was succeeded by 'Dickie' Shaw.

There were two entries in Shaw's 1917 log book that had fascinated Grosvenor, and which he would frequently ask to see. One read, 'Have today practised Rising and Falling'; the other, 'Hit a tree on take-off – flying very much improved.' Shaw nevertheless was the most experienced, and one of the best loved members of 601. He had won his DFC in the First World War on the first squadron to operate anti-shipping patrols from Dunkirk. Flying the frail and spidery DH 4 over water was a hazardous affair, and the number of crews lost through engine or mechanical failure caused the military chiefs concern. They equipped each aircraft with a brace of carrier pigeons to be released in time of distress. The crews of Shaw's squadron were sceptical about the pigeons. For one thing they didn't trust the birds' intelligence, and for another they had better plans for them, for rations were short. When a snap check was made of the pigeons the authorities were appalled to discover a serious inventory deficiency, and orders were issued advising that the pigeons were the property of HM Government, that they were survival equipment, and that any further crew members who abducted them would be court martialled. It was a week before another DH 4 ditched in the English Channel and its crew of two, carrying out their drill, secured messages to the pigeons' legs and released them. The wretched birds circled the floating aeroplane twice, lost heart, and settled on the upraised rudder. As the pilot climbed along the fuselage to shoo them into flight his weight caused the tail to sink into the water, and the recalcitrant birds transferred themselves to the starboard wingtip. While the pilot guarded the tail the gunner crawled along the wing, but as the starboard wing dropped and the port rose out of the water the pigeons changed wingtips, and continued to watch the antics of the two flyers with interest. The crew realised that whichever two extremities of the plane they covered, there would always be a third until their craft sank and it was too late. Luckily they were rescued by a destroyer, but within twenty-four hours an appetising aroma floated over the Dunkirk camp.

The Hon. John William Maxwell ('Max') Aitken owned an Aeronca, G-ADZZ, a deep-chested American high wing monoplane that gave its owner and his friends a great deal of amusement. Owing to its specialised design the plane was claimed impossible to spin by its manufacturers, but after many hours of practice Aitken found a way. An air show was held at Hanworth and Aitken was invited to demonstrate his unspinnable machine. He flew across with Hawtrey, who won the toss and made the demonstration flight with Aitken as passenger. After two ostentatious circuits of the aerodrome at three thousand feet to make sure everyone was watching, Hawtrey put the Aeronca into a spin, did a couple of turns and recovered. Aitken, his broad face creased with laughter, called for another spin before landing and this Hawtrey did, although it was a near thing and they only cleared the trees on the airfield boundary by a few feet. There is no record of what the Aeronca representatives said.

Air shows were popular events and a good way for the Auxiliaries to show regulars what they could do. The regulars were not always comradely. A building containing a bar for the at-home squadrons at one event displayed a sign banning all outsiders. In response an outraged 601 spent the night solidly bricking up its doors and windows.

Enthusiasm for flight knew no bounds, and the pilots never seemed to have enough time in the air. Their privately owned aircraft added to the crashes and mishaps of those belonging to the RAF. Jimmy Raglan, a famous stage and film actor (and on television much later, in the 1950s), would take the milk train from London to Lympne every morning during summer camp, fly all day, and return by the train in time for the evening's performance. He flew a DH 9A into an obstruction light, confirming his friends' assessment that he was a better actor than pilot.

It was held that if a light aeroplane were flown properly there was hardly a field in which it couldn't be landed, and nearly always the squadron was right. During summer camps at Lympne private aircraft were often flown off the aerodrome and put down alongside the favourite pub, Botolph's Bridge Inn, a picturesque tavern among the sheep pastures of Romney Marsh.

On an evening in 1935 the Legionnaires were a assembling at Botolph's before the rabbit hunt. When it became dark, live rabbits would be caught in the headlamps of a motor car driven round Lympne aerodrome, then put through every window found open on the airfield precincts. (Almost a hundred had once been let loose through the window of the Cinque Ports Flying Club, and the barman who found them next morning had to be coaxed into staying with the job.) The evening was warm, and the door of the inn was left open. Through it drifted the strains of a concertina played by Peter Robinson as he leaned against the bridge, and the sounds of aero engines as several little machines rose just beyond the brow of the hills, cruised the two miles then circled over the inn and followed each other into the stamp-sized field opposite. Roger Bushell, a larger-than-life person who was destined to become the Legion's most famous figure, approached in Aitken's Aeronca, made a crisp approach over the taxiing machines, swerved to avoid a sheep and touched down too late. Tearing through the hedge before disintegrated on the highway less than twenty yards from the inn, the aircraft decapitated a lamppost and a sign that read 'To Dymchurch'. Bushell stepped unscratched from the wreckage, apologised to Aitken with a deep bow, and began to auction off the Aeronca's remains to the gathering crowd of spectators. The successful bidder was Mrs Ann Davis of the Cinque Ports Flying club, who paid £5. She sent the Legion a rhyme based on the aircraft's letters:

G one is Max's aeroplane,
A rtful Roger is to blame;
D own at Botolph's Bridge one night,
Z ero was his cruising height –
Z oomed too low – out went the light!

The 'To Dymchurch' sign became one of the squadron's mementos, locked up and displayed with the silver on guest nights at town headquarters. Seeing it on the table one night, an officer asked, 'Did that experience ever make you think, Roger, that it was a hell of a long way to go for a drink anyway?' 'Not really', Bushell replied, 'it only worked out at about three yards to the pint.'

When Shaw retired in January 1936 it was Thynne's turn as CO. Thynne bought his own aircraft when he began to make occasional trips to inspect his firm's branch in Egypt, and since his firm paid the equivalent of first-class boat and train fares and he could fly for less, he saved money into the bargain. On these flights he carried a camera in the cockpit at the request of Air Ministry and, in the rôle of secret agent, took a score of aerial photographs of foreign countries that were useful to British Intelligence during the Second World War. Once he crashed in the heart of Uganda and walked for a fortnight before finding a settlement large enough to furnish him with transport to civilisation. Another time, flying from Egypt, he landed over the Turkish border to refuel. He managed to keep the film, although he was suspected of spying, was told his papers were not in order and made to spend eleven days in a small, beetle-infested cell in Adana before being allowed to continue to England but without his aircraft.

On one trip Thynne met the legendary 'Batchy' Atcherley of RAF, who was flying a light aircraft in the opposite direction. They discussed the problems of navigation over dinner. Atcherley had overcome the major difficulty of finding his way about the Middle East, which was devoid of significant landmarks. 'As a matter of fact', he confessed, 'I fly by Bradshaw [publisher of a railway guide] and follow the rails. I usually borrow a railway timetable wherever I'm staying overnight, and the chances are it'll have a rail map on the back. I cut this out and add it my collection. It usually gets me through.'

Such formative flying was done at a time when no pilot could have hoped to make a long distance flight without the assistance of Shell. Weeks in advance the traveller sent an estimate of the petrol needed – thirty gallons at Marseilles, perhaps another twenty at Tobruk – and the company must have spent a great deal subsidising light aviation, for the fuel, at normal price, would always be waiting in four-gallon cans addressed to the pilot.

Thynne was well equipped to become the leader of a group of pilots as uninhibited and unruly as himself. His command led to a colourful period in the squadron's pre-war history at a time when the men who composed it were strangely insulated from the powerful social and political changes around them. It was the age of aviation, jazz, economic depression, and the Spanish Civil War. It was also the age of growing fascism in Europe and the beginnings of British preparation for possible war. Unconsciously Tory, they preferred to live their philosophy than to think about it. Nevertheless, there was a conviction among those who had been to Germany that things couldn't be like this for long. The 1930s were exactly right for Thynne and the Legion.

Although he maintained the Legion's traditional perversity, Thynne brought his own interpretation of how it should be run. He enforced only two rules: one, every new member must discover as quickly as possible which regulations were traditionally broken, such as dress, for socks of squadron red were worn with uniform, and which observed: for example, it was said, it was forbidden to throw coal at the commanding officer since this annoyed the caretaker. The second rule was no taking off or landing on the tarmac, although Thynne himself frequently did.

A new officer who was at a loss to divine the amorphous, meandering line separating pliable rules that counted from foggy ones that didn't, had the dilemma solved for him by Thynne when he recklessly ventured that, 'Rules are, after all, for the obedience of fools and the guidance of wise men.' 'Exactly so,' agreed Thynne, 'so kindly remember to obey them in future.'

Thynne, like his predecessors, judged it prudent that no person outside the squadron should ever have the slightest inkling how it was run. 'They're quite extraordinary,' a senior officer was heard to say of them, 'very keen and efficient, but they can't have much discipline; do you know, they call their CO Brian!' 'How do you maintain discipline?' one regular had asked Hawtrey, the regular adjutant, to which Hawtrey replied, 'I just say "You mutinous scum!" It's easy, and they all tremble!'

The aircraft identification letters, preceded by the UF allocated to 601, were arranged in contrary direction backwards from Z. The Red and Yellow sections of 'A' and 'B' Flights, standardised for the RAF, were absent, but instead the propeller spinners and other parts of the aircraft were painted in the dark and light blues of Oxford and Cambridge. Only the initiated knew that Oxford blue meant 'A' Flight and Cambridge blue 'B' Flight.

The new Auxiliary force wasn't too quick to shed all vestiges of its origins; even the AAF uniform included riding breeches and puttees (except for the CO, who wore cavalry boots), which seemed natural enough. Each flying day began with fifteen minutes of drill, which would sometimes descend into farce with Loel Guinness giving infantry orders from the Guards, Peter DuCare naval orders, and Thynne cavalry orders from the Yeomanry. The foxhunting cry of 'tally-ho!' meaning fox sighted and of old French origin, was in use throughout the war to mean enemy aircraft sighted. More important were the vague but firmly entrenched Yeomanry ideals of duty, manners, patriotism, and application.

Under his command, in the loosest sense, Thynne was lucky enough to have at one time five university blues and seven international skiers. A natural progression from sports, skis, bobsleighs, and racing cars to flying was strongly suggested by the talent of these young pilots. Danger was a part of it. Not that anybody wanted to get killed, but having the skill and confidence to avoid doing so was deeply satisfying. Imaginative and rarely cautious, they were happiest when pushing their aircraft and their judgment to the limit.

Gordon Neil Spencer 'Mouse' Cleaver had skied for Britain, while 'Paddy' Green broke his leg in the President's Cup at St Moritz in 1937 and was back in the snow again within a fortnight. One star of the slopes was Peter Beverley Robinson, the Canadian son of a League of Nations secretary in Geneva, who was practically brought up on skis and whose success was only marred – in his own eyes – by consistent mediocrity as an actor. He used to do the Flying Kilometre, which the others didn't, and it was the thrill of those long seconds in the air with arms outstretched which, as a youngster, gave him an idea. With a pair of wings strapped to his shoulders he could be the fastest and longest-range ski jumper in the world, sailing effortlessly over villages and forests. Robinson made his wings and, using his body as a fuselage, launched himself from a small alp. There wasn't enough lift, so he made a larger pair of wings and tried a higher alp. He took off immediately but, with some alarm, found he had no control at all, and wisely released himself at thirty feet to fall unharmed into a pine copse. Peter Robinson (not Michael Lister Robinson, the son of Sir Roy, later Lord, Robinson who as a regular in the RAF also joined 601) would shortly leave for the stage in New York.

The Hon. John William Maxwell 'Max' Aitken, twenty-six-year-old son of Lord Beaverbrook and later general manager of the Sunday Express, was a Cambridge soccer blue and scratch golfer. Rugged and handsome, with his father's broad and genial features, Aitken had flown hundreds of hours over Europe and the United States and broken a string of records for transport planes in America with one of Lord Beaverbrook's machines. His powerful Fleet Street connections provided the Legion with a valuable ally.

William Pancoast 'Billy' or 'Little Billy' Clyde came from a family of great wealth, his grandfather having founded the Clyde Shipping Company in New York. He went to Eton and Oxford and became a member of Boodle's, a gentlemen's club in St James's Street just one block south of White's on the same side. All his spare time Clyde spent skiing in Switzerland where he eventually ranked fifth in the world for slalom. Lightly built, he tried his hand as a steeplechase jockey and served as *Aide de Camp* to the Governor of the Bahamas before becoming a working director of the pharmaceutical company Johnson & Johnson in New York. He lived in Princeton, New Jersey and married Rosie.

Roger Joyce Bushell was one of the more intriguing of Thynne's officers, a skiing champion born in South Africa to British parents, with a run named after him in St Moritz in recognition of the fastest time on it. Flamboyant, dashing and tough, he was fluent in German, French and capable in Afrikaans, and could swear in several languages. He could spit an impressive distance. It would be redundant to say he was acutely intelligent. His mother noted when he was young that Roger had the qualities of courage and truthfulness, and here she was half right; courage is easily defined and Bushell's was limitless, but the truth can often be seen from many angles and Bushell was adept at choosing the one which served his professional purposes and, later, his survival.

Wellington College, Berkshire, and Pembroke College, Cambridge, failed to humble Bushell, but in a letter to his mother the housemaster at Wellington had written, unfazed by the barbaric nature of British public schools at the time, 'Don't worry about him. He has already organised the other boys. I know the type well; he will be beaten fairly often, but he will be well liked.' Wellington toughened him. He quickly adapted to England and its easy upper-class lifestyle. Visits home convinced him, with his unbounded self-confidence, that he could become a big fish in a big pond, and with the help of Max Aitken and young Lord Knebworth, both of whom he met at Pembroke, he gravitated to the fringe of British society. It was Knebworth who persuaded him to join The Millionaires' Squadron, after which Bushell invited Aitken.

Dark haired, thickset and of medium height, Bushell caught the tip of a ski under his left eye when he took a tumble during an international ski race in Canada. When it was sewn up he was left with a gash at the corner of his eye which pulled the lid down permanently. Although this didn't spoil his good looks, it gave his face a saturnine quality that belied his characteristic good humour. He possessed a mix of hyper confidence, disrespect for misplaced authority, and social graces, which allowed him to swim in all waters. With engaging eloquence or profanity he could charm a jury or swear with the abandonment of a sailor. Not that he needed polish – his charm was magnetic and universal, his fame spread through the Auxiliary and regular air forces. A station commander, having seen him drink, remarked, 'He looks the sort of chap who, if you turned a [water] tap on, would run away.'

He proved to be an outstanding pilot, described by Thynne as 'the almost perfect flight commander ... and his pilotage exceptional from the start, and in this squadron where everyone was enthusiastic and the normal level was so high to be outstanding was a difficult feat'.

On 601 Bushell lived explosively, happily. He was in his element. He had many girlfriends, some not even married. The Line Book records that 'a scarlet ear ring was found in Roger Bushell's service aircraft. On learning of this the AOC said "I suppose he swallowed the other one"'.

When the squadron was converted to Demons equipped with radio-telephone (R/T), Bushell's forceful language shook the wireless amateurs for miles around London. Once he gave coarse instructions to Red Two to close in, and the squadron received a letter addressed to 'Grosvenor Leader, Hendon Aerodrome'. It explained that by watching them flying and listening to them on his short wave radio the writer had formed a great admiration for the squadron, 'But I am sorry to say that when Red Leader comes on the air I have to send the children out of the room.' This letter brought the pilots immeasurable joy. Thereafter, whenever one of them felt like using strong language he would call up, 'If Mr X is listening he had better send the children out of the room now!' Until it expired from overuse, the expression 'Oh, send the children out' served as a popular expletive.

There was an almost perpetual parallelism in the careers of Bushell and Michael (Mike) Fitzwilliam Peacock, who met in Wellington. They had been born within a year of each other, both in South Africa. Peacock, debonair and popular, was to captain Brasenose College Rugby team at Oxford, and ski for Britain. They became inseparable friends and both moved comfortably from Wellington to university, then to the Bar, where as barristers they shared chambers in Lincoln's Inn, and finally to No. 601 Squadron. They also shared a flat in Tite Street, and almost everything else according to a neighbour, who said that if one went out in tails the other would stay at home.

The services of both these quite brilliant pleaders were freely at the disposal of anyone, regular or Auxiliary, who ran afoul of the authorities. If an airman was charged with failing to secure bombs properly beneath an aircraft, or a pilot court-martialled for stunting over his girl friend's house, one of them would, if asked, act as counsel for the defence. Having studied RAF law they won more than their share of cases and became something of a legend. The authorities were continually embarrassed.

Bushell enjoyed the challenge and theatre of RAF trials and brought to them an unusual flourish which the court often found baffling and intimidating. Once briefed to defend a fitter accused of negligence, he flew a squadron aircraft to RAF North Cotes in uniform but appeared at the court in full regalia of gown and wig, as though in the High Court, impressive and disguised. Alternating between wit and calculated pomp, quoting sources and precedents, he finally drove the key prosecution witness to anger – a favourite ruse – and then demolished him with polite ease, before changing back into uniform and taking tea in the mess. The only other occupant of the anteroom was the humiliated prosecution witness, a much senior officer. They chatted, and to Bushell's delight the other officer was still smouldering: 'Just had a damn rough time at a court martial. Everything was bloody well subverted by some young charlatan of a lawyer from London. Confounded fellow was infernally rude to me, then nobody could understand what he was talking about, not even the president, and they had to let the man off. Outside that court I'd like to tell that scoundrel what I think of him.' Bushell nodded, excused himself, and flew anonymously away.

Mike Peacock used the costume trick when he flew in one of the squadron's new Hawker Demons, a fighter version of the Hart, to defend an officer at Kenley, climbing out of one of Britain's latest fighters in a barrister's gown, having changed his flying helmet for a wig while taxiing in to dispersal. The ground crew were too bewildered to take their eyes off him as he sauntered away.

Perhaps, as with many legal minds, both Bushell and Peacock saw law as a conflict. The air force saw law as discipline. In challenging this concept with such verve and repeated success they strained the patience of authority to breaking point. The last straw was Peacock's defence of a regular pilot on a low-flying charge at Biggin Hill, of which the RAF took a serious view, particularly as complaints had been lodged by some very respectable citizens. Peacock knew that in such cases there was an inclination to attach undue credence to the testimony of the upper and professional classes, but he also knew that with the right tactic it could be difficult to make a low-flying charge stick because perceptions of height were almost always inexact and difficult to prove. With the wiliness of a barrister and the confidence of a man with no aspirations to a career in the air force, Peacock crafted an artful defence.

There were two witnesses for the prosecution but none for the defence because one can't prove a negative. Biggin Hill, nicknamed Biggin on the Bump, stands on a ridge, and the first witness was a vicar who had seen the aircraft flying low in the low area outside its borders. How low was the aircraft, Peacock asked the vicar. Oh, he didn't think he could say with accuracy.

'Was it below a thousand feet?'

'Oh, that I don't know.'

'Perhaps below five hundred feet?'

'I'm afraid I can't express the aircraft's height in such specific terms.'

'Of course, I do apologise. I shouldn't expect you to be familiar with heights expressed in terms of feet,' purred Peacock. 'I wonder, then, whether you might be able to relate the aircraft's height to something with which you are familiar. Let us take – your church spire, for instance. Would you say it was as low as that?'

'Yes, I would say it was'.

Peacock thanked the vicar, then cross-examined the second prosecution witness, an officer who as duty pilot at Biggin Hill was responsible for observing take-offs and landings from the duty hut.

'Now you have testified, have you not, that you observed this aircraft low flying at the time, from beginning to end?'

'That's correct', the officer agreed with obvious discomfort.

'And you have testified to that under oath?'

'Yes.'

Peacock dismissed the duty officer and propped up a prepared drawing that showed that the church spire was hidden on the valley from the duty hut on the high ground of Biggin Hill. He drew a line from each witness's position and stood back, letting it sink in.

All the air seemed to be sucked out of the courtroom as the court studied the drawing. Slowly it dawned on them: the plane couldn't possibly have been both visible from Biggin Hill and below the church spire in the valley.

'Somebody', Peacock quietly said, 'is not telling the truth.'

It was impossible to convict. Air Ministry was furious. Thynne received a message ordering that 'In future, on no account are Flight Lieutenants Bushell or Peacock to act as defence counsel in courts martial.' Of course, anyone choosing to rebuke Brian Thynne had better be sure of his ground, and Thynne realised the order was ill conceived. He replied, enquiring politely whether the instruction wasn't perhaps an error, in which event it might be possible to withhold knowledge of the matter from the national press. Air Ministry quickly retracted this 'mistake'.

Sassoon, on relinquishing command and becoming Honorary Air Commodore, saw the beginnings of a protracted war against 600. It became increasingly frequent for his dinner parties to end prematurely and in wild disorder with the sudden appearance of a dishevelled Legionnaire who had been left on duty at the camp shouting, 'Six Hundred raiding! Six Hundred raiding!' When this happened the squadron guests would be out of the room in seconds. These abrupt and violent departures caused him some consternation, and in due course he would attach to his invitations the condition that, 'You must promise not to leave in the middle of dinner, or in a hurry.' Just to be safe he did what his normally liberal nature wouldn't have permitted and ordered his staff to lock away all breakable works of art when inviting the squadron to dine in any strength.

Both 600 and 601 always believed it to be their God-given right, their duty almost, to make off-duty life in camp as difficult for each other as their inventiveness would permit. At one summer camp that they shared at Lympne, 601 conceived a brilliant insult by mocking 600's emblem, 'The City of London Arms, Overflown by an Eagle', a somewhat complicated design depicting an eagle atop a George Cross, by planting outside their mess a wooden sign painted by the squadron's artist, Huseph Riddle, of a plucked chicken with a flying sword stuck up its rear end. At another camp at Lympne, both squadrons flew the comparatively slow DH 9As, and since no night flying was then undertaken with service aircraft the Air Ministry allowed Romney farmers to graze their sheep on the airfield from dusk to dawn; in fact, this served to keep the grass down. While 600 were at a party in Folkestone, the 601 pilots rounded up thirty or forty sheep and herded them into the 600 tents. In retribution, 600 waited for the last lamp to go out in the 601 tents, then carefully removed guy ropes and pegs and wrapped each sleeper in a mess of damp canvas.

From this small beginning the war escalated, or descended, into a spiral of near lunacy that made the headlines of national papers. Flight Lieutenant Campbell-Orde of 600 flew across to Lympne in a service Hart from nearby Hawkinge, where his squadron was encamped in 1936. He was a good friend of the Legion, but he was also from 600, and that made all the difference. They laced Campbell-Orde's beer with liquor and when he was beyond caring trussed him with old rope and stuffed his tunic and pockets with rotten fish. His aircraft they bedizened with insulting notices that read: 'To and From the Boat Race— 5/-' and 'This Way to the Zoo'. The victim was transported to Hawkinge in a private cabin plane and dumped in the centre of the field.

Predictably, 600's reaction was violent and almost immediate. Before five the next morning they had every serviceable private aircraft in the air over

Lympne, and systematically bombarded the tents with every imaginable obnoxious missile. One by one, and without respite for a quarter of an hour, the 600 pilots dropped their repulsive cargo of dead rabbits, birds, crabs, bombs of soot and flour, and balloons containing a mixture of ink and milk. They smothered the tents with red and yellow ochre, sacks of bad eggs, old fruit, fish, and potato peelings. Finally, as the half-sleeping Legionnaires staggered choking from their tents, cartons of treacle and silk stockings loaded with cow manure fell upon their heads. During the raid, a 600 task force that had arrived by road broke into the mess tent and, amidst the fearful confusion, made off with the much prized wooden Flying Sword.

The commotion was so great that news of the raid broke in the evening papers. But there was more to write about before the week was through.

Nobody, much less 600, could have expected the Legion to endure such humiliation passively, and Thynne flogged his officers for a worthy plan of revenge. Raymond Davis thought of one. No. 600 were to be raided after the fashion they themselves had set, but with two diabolical refinements. First, the raid would take place just before their full dress dinner on Friday night, thus ruining their dinner. Second, a smoke screen would be laid by air to blot out the entire camp while a gas mask party retrieved the Flying Sword.

There was a technical difficulty; nobody had ever produced a smoke screen before. But it was understood that tannic acid would form a dense smoke on contact with the air, so the first step was to try to get some. Thynne set off with Raymond Davis, Peter Robinson and Sir Archibald ('Archie') Hope to an armaments establishment at Eastchurch. They were at first unable to persuade the office in charge to let them have any. 'What on earth do you want it for, anyway?' he asked.

'We want it', said Thynne blankly, 'to drop on 600 Squadron.'

'I see', replied the stores officer, Flight Lieutenant Murray Payne, 'and how much will you need?'

Payne had been demoted as a result of an incident involving 600 Squadron, against whom he bore an unreasonable grudge. He had been in charge of an air-to-sea firing range, and gone to lunch without pulling in the 'Clear to Fire' markers, with the tragic result that two No. 600 Squadron Harts had machine gunned a boat and injured a girl.

The next problem was to find a way of exposing the acid to air at the right moment. This was solved with the bulk purchase of flimsy cardboard ice cream cartons, which would burst on contact with the ground. By Wednesday trials were complete, and on Friday evening Thynne led the mixed raiding force of twelve private aircraft from Lympne, out to sea round

Folkestone, and crossing the coast near Hawkinge. Timing was perfect: the 600 officers, in full evening kit, were drinking their sherry. There was chaos as the tents filled with foul-smelling white smoke, through which firecrackers and the more conventional garbage rained. The smoke screen was entirely successful. So was the gas mask party.

In reporting the raid, some papers hinted darkly that its perpetrators would be 'for it', although the overall press reaction was one of hilarity. Surely it showed the potency of air power. The *Daily Mirror* in sublime ignorance reported that it understood 'the attack was carried out without the knowledge of the commanding officer of No. 601 Squadron' and that 'an enquiry into the unofficial raid will be made by officials of the Air Ministry tomorrow'.

The second statement was true. The 'enquiry' didn't take very long. The Legion had been using private aircraft, and that satisfied the Air Ministry. But the increasingly scatter-brained behaviour had to stop, and the AOC Jackie Baldwin (later AVM Sir John) rather wisely summoned both commanding officers to say that, in his opinion, the contest was a perfect draw and couldn't it be left at that? Both COs were somewhat relieved to leave it at that. Already the soles had dropped off each Legionnaire's shoes since the acid, which they had used neat, should have been diluted with ten parts of water. The 600 Squadron mess kits looked as though they had been attacked by a thousand hungry moths. Nearly thirty tents had been ruined and were costing the officers good money to replace. Some plants belonging to the Folkestone Corporation, planted in gaily painted boxes and intended for decorating the promenade, were reported ruined.

There was such an outcry by the residents of Folkestone about the plants that Thynne decided to settle the matter at whatever cost. It was a squadron custom for Willie Rhodes-Moorehouse to pay for all 'excess' damage, and he agreed on this occasion to finance the reestablishment of good civic relations. Thynne, his pocket thick with notes, outlined his mission to the head gardener and then, with notions that the bill might be anything from £50 to £100 to judge by some reports, opened tentatively by suggesting £5 (around £250 today).

'Why yes, thank you', rejoined the head gardener warmly, 'I'm sure that will cover it nicely, sir'.

Aidan Crawley, a tall, athletic journalist with the *Daily Mail* and a famous Oxford University and Kent batsman, who had recently joined the Legion, nearly died while practising long-distance navigation on one of his flights, but won for Guy Branch, a new pilot who was quiet and slight of build and who was navigating from the rear cockpit, the squadron's first decoration. They set off in a Demon and while over South Wales in worsening weather their main petrol tank ran dry. They began to descend through cloud. Crawley was able to restart the motor on the reserve tank as they broke cloud over the sea, and succeeded in landing at Netheravon. After refuelling, and a little over confidently, they took off again in bad visibility intending to pick up the railway that ran alongside the field and follow it through a parting in the clouds. Crawley missed the railway on his first circuit and, straining to find it on his second, nearly hit one of the hangars. As he slammed open the throttle he accidentally fumbled the adjacent mixture control lever, cutting out the engine, and although they missed the hangar their machine ended up in a blazing heap in a field. Branch cleared himself from the rear cockpit of the wreckage but as Crawley was trapped Branch jumped back into the flames and pulled him out. For this he was awarded the Empire Gallantry Medal.

Grosvenor had unwittingly bequeathed a problem to his successors. The Flying Sword icon, so aesthetic and fitting in itself, became such a symbol of the unity and incredible spirit which persisted after Grosvenor, that Brian Thynne, Max Aitken, Eddie Ward, Roger Bushell and many other Legionnaires, past, present and future, had or would have it tattooed on their arm or chest. The Flying Sword was central in the RAF's standard crest, a blue circle topped by the royal crown. But Grosvenor had omitted a motto, and so for years the 601 crest had lacked the usual scroll underneath bearing some Latin tag which was the standard appendage. Then in the mid 1930s, the Air Ministry decided this wouldn't do, and that the 601's crest should be brought into conformity with all the others. Thynne was requested to submit a motto.

For once Thynne was at a loss. He wrote to several schoolmasters for suggestions and studied works on heraldry, but could unearth nothing that was even remotely apt. There was a dilemma, possibly one Grosvenor had faced: how do you avoid incomprehensible Latin doggerel or sheer drivel? Thynne was requested to submit a motto to the Royal College of Heralds. He wrote to the College to say that, since exhaustive enquiry had failed to produce anything better, he would be happy to continue without a motto. The College replied that this was unthinkable; a plain ribbon under the crest would be ridiculous, while if there were no ribbon at all the entire balance of the crest would be upset. 'Very well', wrote Thynne, 'please design a strip of red tape to fit under the Flying Sword, symbolising without words the frustration of an Auxiliary squadron.' The College's reply ignored this innuendo but pointed out with transparent satisfaction: first, that the design submitted was itself heraldically wrong in thirteen different respects; and second, that if it was to put this right the fee for approving a revised design was ten guineas. This stumped the squadron for a while, but the more they thought about it the more they were convinced that embellishment of any kind would detract from the symbol, and that any wording would be either fatuous, or incomprehensible Latin doggerel. Perhaps this is what Grosvenor had thought. But one foggy, unflyable day, Thynne was rummaging through old files in his office when he came across a copy of Grosvenor's letter to the College specifically referring to the design he had submitted and the College had approved. Pinned to it was a receipt for the fee of five guineas.

This was all he needed. In an icily courteous letter he stated that he had always assumed the evolution of heraldic principles to be leisurely, and the attitude to change conservative. He was therefore astonished to learn that in the ten years since Lord Edward Grosvenor had submitted his design to the Royal College of Heralds to be perfected it had become heraldically wrong in thirteen ways.

The College's answer was contrite and friendly. The Flying Sword design was quite acceptable, and under the circumstances nobody would expect the squadron to pay twice for the College's services. There was no mention of the motto. The badge remained ribbonless, and never again was any attempt made to thrust a motto upon 601.

The Legion in the 1930s was a mix of individuals who made their unique contribution to its character. To capture its flavour it helps to consider Rupert Belleville. One of Grosvenor's recruits from White's, his name was among those that first gave Grosvenor the idea he was forming a Legion. In a collection of colourful characters Belleville stood out. Tall, with thinning fair hair, he was an exceptionally talented, albeit ungovernable aviator, with a persona so famed that a letter from abroad addressed simply to 'Rupert Belleville, RAF' found him without complication. He was intense and mercurial, an adventurer, companion of Ernest Hemingway in Spain, and something of a puzzle to his comrades. He wanted to try everything, and practised as a matador in Spain, appropriately garbed, in a real arena with real flag and a real bull. He was fined £10 plus costs for smoking in a gas-filled airship. He became a pilot for and companion to Mrs Venetia Stanley Montagu.

Venetia Montagu had been a dazzling society beauty distantly related to Drogo Montagu through the web and weft of Britain's establishment. Ardently pursued by Prime Minister Asquith, who would even write notes to her during cabinet meetings, she married Edwin Samuel Montagu, a Liberal Member of Parliament and later Secretary of State for India, and to do so converted to Judaism. Her marriage didn't work well and her affairs included one with Lord Beaverbrook. Montagu died in 1924 and Mrs Montagu took up a keen interest in aircraft and flying, often talking and writing as though she were a pilot whereas she didn't have a pilot's certificate and was always accompanied by someone who did. Belleville (described in the press as 'an amateur flyer') took off from Heston in 1931 with Mrs Montagu, now forty-four, in a DH Gypsy Moth, one of her six Moths, on a six thousand mile adventure across Russia, the Middle East and Persia, a remarkable itinerary given Russia's secrecy over its aerodrome dispositions. These were huge distances for a Gipsy Moth, a wood and fabric biplane with a cruising speed under a hundred miles per hour. As she explained, 'We are going for fun only, in the simplest, cheapest, and most modern way of seeing the world,' which hardly proved the case after they crashed and caught fire in Sabzawar on the way from Teheran to Moscow, though both were unhurt, so that she had to buy another Moth second-hand in Iraq in order for them to continue their odyssey. (Some press reports claimed Mrs Montagu had a replacement Moth sent from her now-depleted fleet of five at home. Either way, cheap doesn't seem to apply.)

When the Spanish Civil War broke out in 1936 Belleville joined General Franco's Nationalist army with a view to getting material for newspaper articles, but soon proved too much for his superiors and was forced to resign. Nothing daunted, with Mrs Montagu's permission he picked up her Leopard Moth at Lympne and returned to Spain as a freelance correspondent. At Sonny's Bar in Biarritz he met Virginia Cowles, an American journalist awaiting her visa for Nationalist Spain, which she feared would never arrive. The svelte Miss Cowles, every bit as adventurous and formidable as Belleville, wrote respected articles for both sides of the Atlantic, had interviewed Mussolini and had tea with Hitler, and acquired a strong affection for Britain.

Belleville told her to forget about the visa and persuaded her to fly with him to San Sebastian in the morning as he was confident that the authorities knew him well enough to grant them entry. Belleville must have been very persuasive. She accepted his offer and, with an English businessman making a third, they took off from Biarritz aerodrome the next morning. To avoid anti-aircraft batteries, Belleville hugged the mountains that ringed San Sebastian to land. But instead of the friendly reception he had promised, all three of them were unceremoniously arrested and their aircraft impounded.

The Spaniards allowed their prisoners to stay at the Maria Cristina Hotel under house arrest, and after a few days gave permission for them to return to France. Belleville, however, had by now begun to enjoy the stimulating company of several bullfighters and Ernest Hemmingway, whom he had befriended, in Chicote's Bar, and was angling for an opportunity to try his hand in a real bull ring, at which, remarkably, he eventually succeeded. In any case he had no intention of leaving without his aeroplane. When the Moth was restored to him he waited around for an opportunity to move forward into the fighting area and obtain his story.

The fall of Santander to Franco was considered imminent. Seeing the local postman running through the street outside Chicote's Bar proclaiming its capture, Belleville, without bothering to confirm this piece of news (it was false), set off immediately for Santander in the Moth accompanied by a young Nationalist Spaniard, Ricardo Gonzales. They were arrested by Republican soldiers while unloading a case of champagne from the plane and shouting 'Viva Franco!' Belleville was forced at pistol point to fly two Republic officers to Gijon, where he was put in gaol. After weeks of negotiation with the British Embassy, Belleville's release was granted in exchange for British promises to the governor of Gijon that efforts would be made to secure the release of an important Red prisoner held by Franco. A British destroyer collected Belleville, but he had to forfeit Mrs Montagu's Moth.

After Belleville's escape from Spain the British government was asked in Parliament, 'In what circumstances, on whose authority and at what cost was a British destroyer dispatched to rescue from the hands of the Spanish Government Mr Rupert Belleville, who appears to have been previously assisting the insurgent forces in Spain; and whether similar rescues have been effected on the other side?' The answer given was that the captain of 6th Flotilla on board HMS *Keith* at St Jean de Luz was informed by a representative of the British Embassy at Hendaye that permission for the release of Mr Belleville had been given and that a motor boat brought him out from Gijon at noon on the following day. 'Arrangements were accordingly made for His Majesty's Ship "*Foresight*", which was on patrol off Gijon and Aviles, to take him on board.' This was done since a ship had been sent 'in accordance with her prearranged programme. No additional cost was therefore incurred. There was thus no question of rescue in this case, and the last part of the question doesn't therefore arise.'

Virginia Cowles saw Belleville again when he had returned to Biarritz. He had just received £400 from the *Evening Standard* for an exclusive story and they went together to the casino where he quickly lost it all. (After the war, Virginia Cowles would marry Aidan Crawley.)

The squadron's Line Book was filled with newspaper cuttings, photos, and handwritten notes mocking anyone who tried to shoot a line, or show off, or who opened himself to ridicule, and a glance through it suggests a harmonious band of accomplished but mischievous young men who didn't take themselves or others too seriously. Entries were made in the book during annual summer camps when training was continuous. The first is dated 2 July 1933:

'I am really a natural pilot.' Roger Bushell

'If you *must* bring SCREAMING harlots into camp take them round to the Cinque Ports Club which is a sink anyway.' The Adjutant

'Nothing is more distressing if you are a continual undershooter than to have a slow ticking prop which may stop.' Nigel Seely

'I can't land a Hart with an instructor. When flying solo I land automatically like a bird.' Tony Gray

'Do you really want me to stop any dangerous flying, because I hear they've got a competition among themselves, and one doesn't want to stop that.' Safety Officer at Lydd, on the telephone to the Adjutant

'What is top rudder? I expect I do it naturally but I don't know.' F. O. Gray

'And then I looked at my instruments – you know, one often does.' Roger Bushell

'You're a shit. At least you would be if you weren't the commanding officer.' No attribution.

The last dated entry, 10 July 1938, is Roger Bushell's: 'When asked to do a simple thing: "It's too fucking hot, and vice versa."'

There are fourteen more entries, all undated but from the same summer camp. The last one is a pasted magazine photograph of a seated lady in a ball gown pointing at a gentleman in dinner jacket who is stooping to speak to her and she is saying, according to the caption, 'I say, do you know your fly buttons are all undone?'

The halcyon days were coming to a close, and everybody knew it. There would be no more summer camps for many a year, and no more entries in the Line Book. It would be stored away for safekeeping with the squadron's silver, and few of those named in it would live to leaf through its pages again.

Chapter 4

Ski Boots and Jackboots

> Poor England! Leading her free, careless life from day to day, amid endless good-tempered parliamentary babble, she followed, wondering, along the downward path which led to all she wanted to avoid.
>
> Winston Churchill, *The Gathering Storm*

The curtains that closed on the 'thirties rang down on a world that wouldn't be seen again. Besides the golden age of aviation, they were the decade of awful advance toward war. Germany secretly, and the other nations openly, expanded aircraft production and strengthened their air forces, and in 1934 the British government followed suit with a five-year expansion plan to add forty-one squadrons to the RAF. The warnings of Winston Churchill called out the gathering strength of Germany's secret air corps, officially established under the Goering the following year, 1935. The new Luftwaffe was fast, modern, frightening to those who understood such things.

The government and the Air Ministry, however, hadn't been idle. In the years building towards war it had developed not only two fast, eight-gun, monoplane fighters that closely matched the Luftwaffe's Messerschmitt Me 109 in performance, the Hawker Hurricane and Supermarine Spitfire, but also an effective radar capability for detecting incoming airborne threats. The Germans knew this, but what they failed to understand was that there was a sophisticated system, called Chain Home, linking the various radar stations to control centres that filtered and interpreted incoming signals and directed fighters to intercept 'plots', as blips on the radar screens were called. That was the hardware. As for manpower, a farsighted plan known as the Empire Air Training Scheme (with the unfortunate acronym EATS) was created in anticipation of a shortage of aircrew through 'attrition', the military euphemism for casualties. It was planned to train close to 50,000 aircrew volunteers a year: 22,000 from Great Britain, 13,000 from Canada,

11,000 from Australia and 3,300 from New Zealand. Aircrews were to receive elementary training in various Commonwealth countries, with costs divided between the four governments. Numerous other nationalities joined the plan, later including the United States. It was a firm plan and not just a set of loose commitments, and would eventually make a major contribution to Britain's war effort. Though it would take time for the benefits to flow through, the plan was destined to train half of all RAF aircrew during the war, and in course of time would change for 601 the meaning of 'Legionnaire'.

During the latter 'thirties there was a resigned conviction in many circles that war with Germany couldn't be averted. No. 601 Squadron shared this opinion. Not only were Hitler's barbaric aims and methods increasingly clear, but many of its members had visited Germany, Austria, and Switzerland regularly to ski and there had met Nazi ski troopers in training whose belligerence was unconcealed. Roger Bushell became increasingly impatient, unaware that the government was preparing as fast as it could, and within him stirred a nascent anti-Germanism which was to flourish in due course as the driving force in his life.

In St Moritz Bushell met the famous American William Meade Lindsley 'Billy' Fiske III, an impressive athlete besides being a society figure and multi-millionaire. At twenty-nine years of age Fiske was rich, popular, and the scion of a well-known banking family. He discovered bobsledding at sixteen while studying in France and was a gold medallist in the sport in 1932 at St Moritz. He was as disgusted as Bushell by the Nazis, and when invited to lead the US team at the 1936 Winter Olympics in Garmisch in Bavaria he had declined for this unspoken reason. Now, in 1939, he had just set a Cresta run record. Over drinks Bushell told Fiske the Germans must be stopped; that he was convinced war was coming, and that as an Auxiliary he expected to be called to active service any day. He described the rush he got from flying, the camaraderie of a fighter squadron, and the unique quality of 601. Fiske was intrigued. He couldn't fly, but wished he could. His chance would come.

Another impatient young Legionnaire was Willard Whitney Straight, the American-born millionaire, aeroplane designer, pilot, and international champion racing motorist. The son of Major Willard Dickerman Straight and heiress Dorothy Payne Whitney of the Whitney dynasty, he graduated from Cambridge. He had spent years in Munich studying art, spoke fluent French and German, and understood the contemporary German mind well enough to be certain war would come. After leaving Germany he returned to England and opened up several civil aerodromes. This activity took him back to Berlin for an aerodrome builders' convention, and there he was shocked by the mild exasperation of a senior British Embassy official who complained of 'a tiresome man named Hitler who wants to move the Consulate'. In 1936 Straight became a British citizen. The following year, believing he could not face a future in which he hadn't contributed all he could to the destruction of the Nazis, he joined 601.

The squadron's existence assumed a new sense of purpose as it prepared to demonstrate that despite the horseplay it was a well trained combat unit ready to take its front-line place in the country's air defences. The defence concept of the Auxiliaries' rôle, however, was relatively recent. They had originally been established as bombers, but the increasing multiplicity of bomber aircrew categories and the logical need for a stronger air defence brought about their conversion to fighter interceptors.

The official attitude to the Auxiliaries was interesting. There were signs that somewhere in the higher echelons of the Ministry there persisted elements of unsympathetic opinion. In the early years of their existence the Auxiliaries had been excluded from competition for a trophy that was awarded annually to the most efficient bomber squadron, on the grounds that they were inexperienced. As time went on the revised explanation for exclusion was that 'it would create an unfavourable impression were they to win'. In fact, surprisingly little was understood about them by some people who should have known better. At an Air Ministry conference on Auxiliary capabilities attended by the commanding officers, a civil servant strove to clinch his point by saying, 'Even you must admit that you couldn't have part-time COs.' 'Then what in God's name', thundered the commanding officers, 'do you think we are?'

As prime minister, Stanley Baldwin had expressed little interest in fighters and was famous for his oft-quoted opinion that 'the bomber will always get through', which was doubtless true if he meant at least some bombers, but it led him to a false conclusion. He believed that the best form of defence was offence, thus bombers. But at some point, probably on economic grounds because it was cheaper to manufacture fighters than bombers and they required less aircrew, he threw his support behind the private-enterprise development of the Hawker Hurricane and Supermarine Spitfire, each powered by a twelve-cylinder Rolls-Royce Merlin engine.

In 1937 Baldwin retired as prime minister and was replaced by Neville Chamberlain, a man of peace whose name since his failed negotiations with Hitler has been inseparable from the words 'Munich' and 'appeasement' (though he was squarely behind preparations for war and was the man who eventually declared war on Germany). The RAF continued its program of major expansion. Airfields and accommodation for officers and men were built to a high standard. Social qualifications for aircrew were relaxed, and if you were not a grammar school boy you could at least be a sergeant pilot. (The Luftwaffe was just as stuffy only more so, but on different grounds: you had to be German or Austrian and of Aryan blood though not necessarily a Nazi.)

It was realised that only the course of events could enable the Auxiliary vision of Trenchard, Hoare, Baldwin and Grosvenor to be judged. Each training weekend assumed a new significance. On 27 September 1938, during the Munich crisis, the mechanism of mobilisation worked with gratifying smoothness. A cipher message ordered 601 to move to Biggin Hill and an air party was on its way within an hour, the whole squadron being a complete unit within twenty-four hours of the alert. However, 601 were not unduly disappointed on being ordered back to Hendon on 3 October and disembodied seven days later. Everyone knew that the Demon biplanes had no hope of even catching the German twin-engined bombers. However, the government had begun a crucial shift in thinking away from the RAF's traditional role of bombing towards that of fighter interception. The squadron was pleased, therefore, when its Demons were replaced by faster, albeit still biplane, Gloster Gauntlets. It was a small step in the right direction, and meanwhile the factories were retooling for the new fighters.

There were two other training deficiencies of which the pilots were aware. One was a lack of night flying, which was practically total. The other was inadequate 'live' gunnery. Some firing had been done at towed targets but the greater part had been with camera guns from which theoretical scores were deduced. Both deficiencies were shared by the regulars. More crucial than either of these was a deficiency of which no one seemed to be aware: tactics. Throughout the RAF the approach to aerial combat was hopelessly unrealistic, even foolish. Attacks on bombers were formalised to a series of compulsory pre-staged and numbered patterns with no provision for defence against escorting fighters.

When a signal ordered the camouflaging of Biggin Hill, surplus paint stocks were hurriedly purchased from a London company and applied to the airfield fixtures. The runway and the taxi tracks were decked in garish purple, the hangars orange and yellow, and surrounding buildings blue. Pilots said that from the air nothing could have looked more hideous or conspicuous.

The 'thirties began with cloth-helmeted 601 pilots wearing goggles and flowing scarves in open cockpit aircraft. The squadron had flown such types as the Avro 504, Airco 9A, Westland Wapiti, Hawker Hart, all light bombers, and the Hawker Demon and Gloster Gauntlet, both fighters. In the thirteen years of its existence up to the outbreak of war, therefore, 601 had had six aircraft types, or roughly a different one every two years, with varying roles. Although the Hart introduced clean aerodynamic lines and had excellent performance for its time, all were evolutions of First World War designs, wire braced, fabric covered biplanes with fixed undercarriage and one or two open cockpits, the rear one for the gunner/observer. The Gloster Gauntlet was the first single-seat fighter though still a biplane. At Biggin Hill the squadron now received Bristol Blenheim fighters, its first new-generation monoplane and also its first twin-engined aircraft, carrying a crew of three, with a pod under the nose housing four machine guns said to have been made by the Southern Railway. It was a strange type for a fighter, though an interim, and the turnover of types and rôles could not have done much to build flying and tactical skills.

Training on the Blenheims reached its climax at the August camp of 1939 at Ford, near Tangmere, Sussex. A photograph shows the officers in formal splendour in front of a Blenheim, immaculately turned out, some wearing leather gloves, positioned in perfect symmetry around Brian Thynne, the CO. A model of proud regimental posing, it would be last of its kind; muddy boots would replace shoes, and uniforms would crumple under Mae Wests (inflatable life jackets) and parachute harnesses. One of the pilot officers, Julian Smithers, had still to win earn wings, yet within a year he would be thrown into combat against the world's most experienced air combatants.

The end of Thynne's command had arrived and he proposed to Bushell, the senior flight commander and his potential successor, that he would gradually fade out so that Bushell could run the autumn training programme. The last night at Ford was to mark the transition of command. As always, the mess tent was lit by oil lamps, but that night the light was so dim that the diners could scarcely see their food. Thynne suggested to Willie Rhodes-Moorehouse, the mess secretary, that the lamps hadn't been properly filled. 'My dear Brian', said Rhodes-Moorehouse reproachfully, 'didn't anyone tell you that in your honour we're burning Kummel?' Thynne's uniform was ceremoniously burned after the dinner. A fortnight later, the day after Chamberlain's personal letter to Hitler, telegrams went out to all members, the striped trousers were locked away, and 601 was again embodied at Hendon.

At dawn on 1 September the Germans entered Poland. No. 601 was back at Biggin Hill when war was declared two days later. Biggin was more businesslike than the year before – the camouflage had been rectified, communications improved, security tightened up and the aircraft made capable of shooting. Since the Blenheims had no armour plating, Whitney Straight ordered some armoured seats on his own account from Wilkinson Sword Company and had them installed in the machines. Some plating that he obtained privately from Bristol's proved on trials to be too heavy and had to be removed.

Upon Germany's invasion of Czechoslovakia Bushell wired Billy Clyde in New York and Peter Robinson, then acting on Broadway, to 'come back now if you want to be in the action'. The official summons followed. Clyde quickly told Billy Fiske, 'Now is your chance.' All three promptly closed their affairs, packed their belongings in steamer trunks, and set off from New York for Liverpool on the Cunard's RMS *Aquitania*. They were at sea when on 3 September 1939 Prime Minister Chamberlain's tired voice announced that Britain, having received no reply to his warning against invading Poland, was now at war with Germany.

Clyde and Robinson made straight for their unit at Biggin Hill, but Fiske had RAF selection and training ahead of him. He would later rejoin them at Tangmere. At Biggin Hill many of the wives and girlfriends were in WAAF uniforms, having signed up in order to serve alongside their men. The officers were drinking gin and tonics and some were playing poker for £100 a stake. The casual, friendly atmosphere was as Clyde recalled it years before, but not the aircraft. The last time he had flown was in 1936, and in a Hart, but the Blenheim was a twin-engined monoplane with an enclosed cockpit and retractable undercarriage. Thynne deflected Clyde's offer to jump right in and solo, wisely insisting on having a safety pilot accompany him on the first circuit. Despite the many new controls he had to learn about – 'taps' as Clyde later described them – he had no difficulty and was certified on Blenheims in about an hour.

Bushell, Aitken, and Rhodes-Mooehouse received the recall telegram while in the south of France. Rhodes-Moorehouse was with his wife, Amalia, Aitken with his girlfriend. All three flyers were instructed to report for duty with 601 where – the good news for Bushell – it was confirmed that he would replace Thynne as CO. Bushell returned via the Imperial Airways flying boat *Cambria* from Marseilles to Southampton. Meanwhile, Thynne rushed off to Moss Brothers for a new uniform and reappeared at the head of the Legion to inform Bushell on his return that their succession agreement, like an insurance policy, excluded the exigencies of riot or war. This seems reasonable; no self-respecting commanding officer could allow himself to appear to be turning tail on the brink of battle. Whatever Bushell made of it, and it isn't recorded, his disappointment was eventually allayed by his promotion to squadron leader, charged with re-forming an old fighter squadron and nemesis, No. 92.

Many 601 officers were using motorcycles for petrol economy, and when it was learned that petrol rationing would take effect within a few days the squadron came nearer to panic than it had ever been. Thynne called a hurried meeting at which it was decided to stockpile fuel. First they had to get some. Willie Rhodes-Moorehouse was appointed petrol officer, relieved of all other duties, and told not to come back without results. The next morning he was back at the aerodrome.

'Well,' said Thynne, 'How much have you got?'

'I'd say almost enough to last the war.'

'How on earth did you do that?'

'I've bought a garage'.

The owner of the filling station had been glad to sell. The road it was in, which ran through the camp, had been closed a few days before, and Rhodes-Moorehouse had drawn a cheque on the spot. But the tanks were only half full and his estimate sounded a little optimistic, so the 'soviet' convened again to discuss this problem. A light dawned slowly in Loel Guinness's eye.

'I'm not sure', he said hesitantly, 'but I think I'm a director of Shell'.

'What do you mean, you think you are?' snapped Thynne, 'Go and telephone your secretary and find out!'

Guinness's secretary confirmed that he was on the board of a Shell subsidiary, and within days the tanks of the garage were brim full, a matter of hours before the enforcement of rationing. The squadron was overly liberal with its petrol and stocks were depleted sooner than expected. When the road was reopened, civilians were mystified to find the pumps of the filling station locked and a large notice in black on a yellow background: 'THE MAN WITH THE KEY HAS GONE TO TEA.' The phrase had its origin in pre-war Sunday cross-country flights when pilots would frequently see this sign at RAF airfields.

Immobilised, the pilots sold their motorcycles and Rhodes-Moorehouse sold the garage at a small profit upon the reopening of the road. The Legionnaires sat around the Biggin Hill crew room drinking beer and playing backgammon, or on the airfield grass, eating splendid lunches delivered by Fortnum and Mason. Only occasionally did they remember that they were at war.

The first alarm was naturally dramatic. Twelve Blenheims were hurriedly scrambled to engage enemy intruders in what promised to be the first air battle. But the 'wong, wong' picked up by sound detectors turned out to be from the motor of a nearby refrigerator. As the Blenheims returned to Biggin Hill they were fired on by Kent anti-aircraft gunners but the nearest miss was over a thousand yards.

Gatherings outside the crew room took on the character of family affairs. By the outbreak of war many of the Legion's pilots had married, often to each other's relatives. Archie Hope was married to Raymond Davis's sister and Davis was married to Hope's; Tom Hubbard to Guy Branch's sister; and Henry Cavendish to Rhodes-Moorehouse's first cousin. Drogo Montagu, the son of Lord Sandwich but no longer with 601, had been married in turn to Loel Guinness's and Max Aitken's sisters. 'If this sort of thing goes on much longer', bachelor Roger Bushell had grumbled, 'this squadron will be as inbred as an Austrian village.'

Montagu became an instructor and in 1940 died along with his pupil when their Harvard trainer crashed. A Legionnaire who recalled Montagu's earlier wild episodes, including crashing a Wapiti's trailing aerial into Sassoon's mansion and beating up White's Club in broad daylight, couldn't resist cruelly commenting that, 'Drogo was the only pilot to spin a plane into the ground twice, the first time not being fatal.'

For 'Willie' Rhodes-Moorehouse, son of the Victoria Cross hero of World War One, flying had always been a dream, and he was able to achieve this because the family of his school friend George Cleaver owned a plane. Rhodes-Moorehouse earned his private pilot's licence before leaving Eton, at the age of only seventeen. Handsome, debonair, and a keen sportsman, he was selected for the 1936 British Winter Olympics team, but was unable to compete owing to an accident on the ski jump. He travelled extensively, enjoying the South of France and the ski slopes of St Moritz, where he met and fell in love with Amalia, the sister of Legionnaire Dick Demetriadi and a notable beauty who, it was often said, had declined an offer to screen test for the role of Scarlett O'Hara in *Gone With The Wind.*

Thynne received an Air Ministry letter that prompted him to call a meeting in his tent. 'Gentlemen', he announced, 'I have here a communication which will shake you from your apathy and demonstrate to you that the air force is properly preparing for the severities of war.' He read the letter aloud. It offered, in perambulating language, 'photographic likenesses of their Majesties the King and Queen' which, if accepted, were to be hung prominently and facing each other on opposite walls of the officer's mess. A method of acceptance was prescribed. There the matter would have been left, but next morning Mike Peacock, his face unusually solemn, entered Thynne's office with a piece of paper in his hand.

'Sir', he began with unaccustomed formality, 'knowing how little time you have for these matters, I have taken the liberty of drafting a reply to yesterday's letter from the Ministry.' Thynne read the draft and promptly sent it. In ponderous terms which parodied the Ministry's letter, it expressed gratitude for the offer but explained that questions of such magnitude had always to be resolved by consensus. 'The soviet met yesterday', it went on, 'and instructed me to decline your offer with regret. This course should in no way be construed as disrespectful to the Crown but the contrary, for if the likenesses were accepted it would be necessary to hang them back-to-back, our mess walls consisting at this time of one tent pole.'

For a while the war was little more than an extended summer camp. When orders came through to set up training, and the long hours of 'Readiness' began, the change become apparent. There was the occasional panic during the 'Phoney War' – a scramble passed by telephone, which amounted to nothing, or a briefing for some chore. When an officer in the crew room answered the telephone linked directly with the operations room the atmosphere would be electrified. Sometimes there would be a misunderstanding as the pilots at readiness heard him repeat the information passed: 'What was that – Thames estuary? Yes, and how many are there? Eleven. What altitude? Five thousand ...' So low? The pilots would gather up their kit and feel their pulses quicken. Half way out to their aircraft they would be called back and told that the message was only a warning about barrage balloons in the Gravesend area.

It nevertheless seemed that Mike Peacock would engage the enemy when he was scrambled to investigate a suspicious plot over the Dutch coast. The controller, with mounting concern, instructed him to 'BUSTER! BUSTER!' (go full throttle), and gave successive changes of course with the assurance that he was closing, closing, still closing on the other aircraft. Although visibility was good, Peacock could see no other aircraft in the sky, and it dawned on him that the controller had mistaken one set of signals for two. After some reflection he asked, 'If I succeed in what you evidently expect me to do, which is to fly up on my own chuff, will it earn me a medal?'

Among the newcomers was the first regular pilot, Pilot Officer 'Paddy' Byrne, an unsung hero of the Battle of Barking Creek – an inadvertent dogfight over Essex between Hurricanes and Spitfires from North Weald and Hornchurch in which two of Britain's precious fighters had been lost and one pilot killed. The sardonically nicknamed battle occurred in the early morning of 6 September 1939. Britain's radar system was in its infancy, and when signals emitted by the radar station were echoed and, as with Peacock, appeared to double the size of the supposed aircraft formation approaching from the east at high altitude over the Essex coast. Six Hurricanes from No. 56 Squadron at North Weald were scrambled. In confusion, additional fighters were sent up to intercept, giving controllers at their radar screens the impression of a rapidly growing enemy armada. Quite independently, two pilots took off in reserve Blenheims and followed the interceptors at a distance, and they would become the victims of the mistaken attack. Adding to the confusion, more Hurricanes from North Weald and Spitfires from Hornchurch were scrambled. With the war only three days old, there was as yet no way of distinguishing enemy aircraft from friendly on the radar screens. None of the RAF pilots had seen combat and very few had ever seen a German plane. Communication between planes and controllers was in its infancy. All the pilots expected to see enemy aircraft and the leader of one of the Hornchurch Spitfire squadrons, the subsequently famous 'Sailor' Malan, gave the order to attack the Blenheims. Byrne and Freeborn did so. Freeborn's victim, Pilot Officer Hulton-Harrop, was struck by a bullet in the head and died before his plane crashed at Manor Farm, Sussex. Malan, obviously to deflect blame, claimed to have cancelled his order with a last-minute call of 'friendly aircraft – break away!' but whether this actually happened it is certain that the attackers didn't hear it.

It is impossible not to sympathise with the unfortunate killer Pilot Officer Freeborn. That it was a tragic mistake is obvious. The counter-order wasn't heard by anyone but, incredibly, Malan testified for the prosecution at the court martial and described Byrne as impetuous and irresponsible. Byrne's counsel was the powerful and blunt Sir Patrick Hastings, a politician MP and former Attorney General, and his deputy was Roger Bushell, the lawyer and fighter pilot. Hastings called Malan 'a bare-faced liar'. The court at Fighter Command's headquarters in Bentley Priory completely exonerated both of the Spitfire pilots, ruling the case an unfortunate accident. Bushell, always a friend of the underdog, found a home for Byrne in 601.

Concise training instructions arrived, and practice interceptions, searchlight cooperation and formation flying supplemented the occasional escort flights. Gunnery and night-flying deficiencies were tackled, night flying being conducted from the old Fairey airfield at Heathrow.

The British Expeditionary Force (BEF) landed in France, and two land forces took up positions along the Franco-Belgian border. Reconnaissance planes droned high and unmolested over England and Germany, dropping leaflets but no bombs. The slow start of the air war was a deception, a summoning up of strength, and it seemed that the only way a fighter squadron could meet the enemy would be in an intruder role. Unexpectedly, such an opportunity arose.

A signal ordered six of the squadron's Blenheims to Northolt on 27 November, minus gunners and navigators, and that afternoon Peacock led three from 'A' Flight and Aitken three from 'B' Flight to Northolt where they were assigned Coastal Command navigators and Bomber Command gunners for the sortie. Ground wireless personnel and armourers arrived shortly afterwards and spent the entire night equipping the aircraft for the operation. Six further Blenheims of No. 125 Squadron also arrived to make up a joint effort.

Wing Commander Orlebar, the station commander at Northolt, had been the pre-war captain of a victorious Schneider Trophy racing team. As the crews assembled and lit cigarettes, he began the briefing breezily.

'You're all very lucky, gentlemen. You are about to go to Germany!'

There was a cool silence, a gentle murmur of uneasy laughs, a sarcastic comment or two, then silence again. Since November, explained the station commander, the Germans had been laying magnetic mines in the approaches to East Anglian harbours and in the Thames estuary. Though the degaussing cable had eliminated any real danger, they were a nuisance, bottling up the ports and occupying the sweepers. This nuisance would therefore be dealt with at its source – the German seaplane base at Borkum – tomorrow. No bombs would be carried, but front gun attacks would be made. Squadron Leader Pott was to lead his section of three, followed by the two sections of 601. The weather forecast was poor, but good enough for the mission unless it deteriorated. The attack was timed for last light, and wireless telegraphy (W/T) would be fitted to the aircraft overnight.

There was little attention paid to briefing at this early stage of the war, and it was certainly not the briefing officer's fault that planning appeared cursory. None of the pilots had discovered, or yet been told, the effective range of their machines. They were not advised where or how they were to land back in the total dark of a cloudy November night. Borkum was two hundred and fifty miles across water, yet not one of the crews possessed an inflatable dinghy or even a Mae West, meaning that any crew that came down in the sea was almost certainly doomed. Moreover, Blenheims were notorious for engine failure. Max Aitken suggested that to increase their effective range they might take off for Bircham Newton in the morning, an airfield near the coast, and refuel there. This was agreed to. The officers at Bircham Newton were intensely curious of their visitors, and over lunch in the mess they learned that the destination was Germany. 'In that case', one of resident pilots said, 'there's something in the hangar I'd like to show you.' It was a Whitley, veteran of a raid, colandered from end to end and with a hole in its fuselage the size of a Blenheim's entire tail unit. 'It used to be mine', said the Bircham pilot, 'but the flak got at it. Have a good trip!'

At a quarter past two in the afternoon the twelve Blenheims took off, formed up and headed east. Hope and Davis were in Peacock's section, Rhodes-Moorehouse and Tom Hubbard in Aitken's. All three leaders had more than a thousand hours behind them. Pott's engine began to give him trouble and nearly cut out three times, which given the Blenheim's known tendency for engine failure must have caused him some anxiety, and he called to say that if it misbehaved again he would have to turn back. Whether the engine recovered or not, Pott remained in the lead without mentioning it again. Scudding just beneath the cloud, the aircraft raced across the countryside, a seaside town with a narrow strip of sand, and over the grey North Sea. As the mist grew thicker and rain blurred the windscreens the formation tightened up instinctively. The gunners fired a short test burst, the pilots scanned their instruments intently. The weather got slightly worse.

Darkness was falling as the Blenheims emerged from a rainstorm on to their target. The Germans were taken completely by surprise and resistance was slight. Coastal patrol boats and light flak emplacements opened up, hurling balls of coloured tracer into the sky as the aircraft raced at full throttle below the level of the hangars and cranes. Bullets from their nose guns scribbled patterns in the water of the harbour and straddled seaplanes in the slipways as figures in blue jerseys rushed to man machine guns on the hangar roofs. Peacock flew through a gap in the mole and fired, as he later claimed, upon a 'man in a bowler hat who must have been the German equivalent of a works and bricks foreman'. The man was running across a hangar roof from which, after stopping to retrieve his 'bowler hat' (no doubt a helmet), he was blasted by the rake of Peacock's fire.

It was all over in a matter of seconds, a few brown puffs marking the scene, and the raiders turned for base. Not one had been hit by flak. Aitken, who led the last section in to attack, was supposed to lead the formation out, but his compass wasn't functioning and gathering darkness split the formation into loose groups. The radio transmission equipment, as expected, failed to work and the navigators gave their pilots dead reckoning or calculated courses, which brought them to England, under fire from British convoys and narrowly missing the balloon barrage at Harwich. It was fortunate that the weather in England had improved, and that there was night-flying practice at Debden, Essex. As the pilots searched for a place to land they saw Debden's runway lit by flares with a searchlight flickering from the perimeter, and all twelve Blenheims landed there. Though of minor consequence in the early days of the war and now forgotten, the Borkum raid was not only the first operation of the Legion, but it was also the first time that British fighters had crossed German territory. The DFC Michael Peacock was awarded for this single operation must have been symbolic of this fact, but Peacock simply put it down to shooting the man with the bowler hat.

Sassoon had seen the growing threat to peace as clearly as anyone, but was unable to witness his beloved squadron go to war and prove all that he had claimed for it. Sir Philip Albert Gustave David Sassoon, 3rd Baronet, PC, GBE, CMG, former air minister, died on 3 June 1939 aged fifty and was cremated at Golders Green on the 5th. His ashes were scattered over the grounds of Trent Park from a 601 aircraft, followed by a tidy formation fly-past that would have pleased him.

Chapter 5

Where is the RAF?

Flew five hours before lunch. AOC asked, 'Are you tired?'

Squadron Leader Aitken's Log Book, 29 May 1940

The Germans had for some time been both fascinated and puzzled by 601 Squadron. They saw it as an admirable cohort of patriotic volunteers but also as a bunch of rich and flippant amateurs. In 1928 a photograph appeared in a Berlin glossy magazine of one of Grosvenor's parades outside 54 Kensington Park Road, in which the faces had been left untouched but the uniforms changed by clever artwork into the ironmongery of mediaeval knights. The caption read, 'Lord Edward and his Ironsides.'

Lord Haw Haw in Berlin (William Joyce, hanged as a traitor to Britain after the war), ever eager to broadcast the latest gossip, made frequent and often flattering references to the squadron, which were received with cheers on the mess radio. Early in 1940 he announced accurately that 'the famous 601 Squadron has been posted to Tangmere', on the Sussex coast. An airfield close by a charming village of that name, Tangmere lay seventy miles west of Lympne along the southern coast.

At the same time, the Blenheims were replaced by single-seat Hawker Hurricanes. A generational leap from the sturdy Bulldog it replaced, the Hurricane was fast and tough but slightly inferior to the Messerschmitt Me 109, Bf version, which could fly faster and thanks to a two-stage supercharger higher than the Hurricane and Spitfire and could out-climb and out-dive both of them. The British fighters had smaller turning circles than the Me 109. The Me 109 had either a 20 mm cannon in the wing and machine guns on the engine cowling, or a cannon placed to fire through the propeller hub plus machine guns in the wings. The cannons had a powerful punch but their rate of fire was comparatively slow. The British fighters had eight .303 inch machine guns, four in each wing, firing incendiary, tracer, and standard bullets, harmonised at what proved to be too great a range, spraying bullets like a shot gun and increasing the chances of a hit but diminishing its punch.

Top shooters had their guns harmonised at two hundred yards and got in close for a kill. At this range and with their high rate of fire they had a better chance of hitting an evading target than would a cannon. The Spitfire and Hurricane had more space in the cockpit and better visibility all round through the canopy. The Me 109 had self-sealing petrol tanks, an obvious advantage, and fuel injection allowing negative gravity so that the pilot could push his stick sharply forward into a steep dive without the engine cutting out. But it had a wider turning circle. And so it went on. Moreover, the relative advantages changed with altitude. The differences largely cancelled out or were marginal and in the end it came down to pilot skill and, above all, tactics. Here, regrettably, the Luftwaffe was undeniably superior.

The Hurricane, though less glamorous than the Spitfire, was simpler to manufacture was a steadier gun platform, could take more punishment, and could be turned round much faster between sorties. It was the predominant British fighter throughout the Battle of Britain and naturally accounted for most of the enemy aircraft destroyed.

With other fighter squadrons of No. 11 Group, 601 shared the role of protecting London, the home counties and southern England as far north as Duxford, where it bordered No. 10 Group, and as far west as Exeter. According to Flight Sergeant Gilbert Henry Harnden, who had joined 601 as an aircraftman in 1932 and was now in charge of 'A' Flight maintenance, 'On March 1st we received our first Hurricanes, second hand ones, some even had fixed-pitch airscrews and hadn't received the best of attention, but we were to find them very much easier to maintain than our old shadow factory built Blenheims, yes it was a real relief for the NCOs to get rid of those Blenheims, as in winter time the engines gave us unending trouble with unexplainable cutting out.' The pilots were just as glad to be rid of engines that had a habit of cutting out,

The weather improved, morning mists were followed by sun and light winds, and after a month's flying the Hurricane was mastered. The CO ordered a pint of beer for each airman and reported ready for combat.

At Tangmere many of the pilots rented cottages around the local village of Bosham. Soon they discovered The Ship. This was a pub, no ordinary pub but a smart club owned by Trevor Moorehouse, an old friend of Willie Rhodes-Moorehouse but no relation. Trevor's idea was to have The Ship Club run by his wife, Nanky, to give her something interesting to do. He was in one of the Guards regiments and usually away. Nanky made a great success of it and the club earned the reputation of having the best wine cellar on the south coast. But the plan nearly fell apart after Dunkirk when the threat of invasion resulted in an order that to provide freedom of action for British guns everybody who didn't have to live on the coast must move inland. Thus Nanky lost nearly all her customers, but 601 was happy to fill the vacuum.

Most of the Legionnaires were married and wanted their wives as close as possible. Jack and Molly Riddle, for example, had only been married in May 1939. The women kept together and were able to look after each other in agreeable surroundings. And they were safe. In the event of invasion they would be advised by Tangmere Operations to get in their motor cars and drive northwards as fast as they could. It was also a perfect solution for Nanky. The pub-cum-club became a popular meeting place for friends and the wives. David Niven, a tall and dashing English actor and Hollywood heartthrob, became a personal friend of several of the officers and visited The Ship frequently. It was here that he met his future wife Primula Susan Rollo, the American-born daughter of a British lawyer, who was serving with 601 as a WAAF. It is unclear who Primula's squadron connection was, though it could have been no accident that she was a WAAF, or whether she wore a silver flying sword, but none of that would have mattered to Niven.

The three Tangmere squadrons alternated with the various stages of preparedness and the most relaxed state was 'thirty minutes available'. This allowed the pilots to leave Tangmere provided that they were on the end of a telephone and could if needed return within thirty minutes. The Ship and all the cottages met this requirement. Huseph Riddle dozed off while at The Ship, and when suddenly woken up by a call to readiness was startled to see his toe nails had been painted green. (It isn't known who paid this compliment to the artist.) There was no time to remove the paint. Had Riddle been captured, his comrades speculated happily, he would have been a perfect example of the degenerate order the Nazis were committed to stamping out. Even this minor playfulness hints at an awareness of the enemy's dark nature.

Loel Guinness, next in seniority when Brian Thynne was promoted to a wing commander controller's job, was now the CO. Bushell had left to re-form 92 Spitfire squadron and Mike Peacock, continuing the strange parallelism, was promoted to command 85 Squadron, equipped with Hurricanes. Aidan Crawley was called away to the Balkan Intelligence Service before beginning his eventful combat career, and Whitney Straight, by his own design, was detached by Air Ministry for secret duties.

On the ground the battle was going badly. The Allied line along the Franco-Belgian border yielded to Von Rundstedt and German armoured divisions poured in through the gap between Namur and Sedan. French and Belgian forces, retreating from the Givet Line defending Belgium, fought desperately to avoid encirclement. The mad rush to evacuate the BEF went into its first stage at St Valery, and here the Green Howards were repeatedly dive-bombed by Ju 87 Stukas (*Sturzkampfflugzeug*, 'dive bomber'). At St Valery, Max Aitken shot down a Messerschmitt Me 110 and a Stuka. The Merville flight destroyed six Heinkel He 111s in one engagement without loss. Four 601 pilots were shot down or forced down over France during the day, but none were injured.

Further negotiations between the French and British governments for the continuation of the war were now imperative. In Britain a new coalition government was formed on 10 May under Churchill, and to the new prime minister fell the task of consulting Paul Reynaud, the French premier, in Paris, in hopes of persuading the French to hold out.

An order was received by 601 at Tangmere on 31 May for nine Hurricanes to escort two Flamingos, twin-engined, high-winged former airliners, with VIPs from Warmwell to Paris. Three sections of three Hurricanes flew higher up to forestall any attack. On arrival the VIPs turned out to be Mr Churchill and key members of the War Cabinet. The Prime Minister's first words were a rebuke to his pilots that he hadn't seen his 'Spitfires', as he called them, throughout the trip. (The mistaken idea that escorting fighters should be close by instead of well above was shared by Goering, whose insistence on this cost many a Luftwaffe life.) Rather than ruffle the new Prime Minister with explanations the situation was tactfully remedied on the return flight by placing one of the sections close enough to his Flamingo for the passengers to see the pilots' faces.

The prime minister had hoped to leave Paris at six in the evening, but the tense conference took longer than expected. Frustrated at being within striking distance of Paris in its last days, the pilots held their own conference from which Hope emerged with a proposal to the VIPs. It was getting late and an escort would be useless in the dark. If Mr Churchill didn't appear by seven o'clock his pilots would be in the city; was that all right? A telephone call was put through to Mr Churchill, who replied that he would stay the night and that they were released until eight next morning. The nine Legionnaires took full advantage of their release. One of Billy Clyde's film star friends, Robert Montgomery, then in Paris, lent them some clothes while Archie Hope was able to obtain money from a friend at the American Embassy.

There are two somewhat different accounts of the next morning, the first of which was from Major General Sir Edward Spears, one of the VIPs and a French-speaking confidant of Churchill, in *Assignment to Catastrophe*:

PARIS, Saturday, June 1st, 1940

On the aerodrome I saw a picture and received an impression of beauty unequalled in my life. The nine fighter planes were drawn up in a wide semi-circle around the Prime Minister's Flamingo. Very slight they seemed on their undercarriages, high and slender as Mosquitoes. Churchill walked towards the machines, grinning, waving his stick, saying a word or two to each pilot as he went from one to the other, and, as I watched their faces light up and smile in answer to his, I thought they looked like the angels of my childhood.

As far back as I can remember I have been enthralled by pictures of angels; Michelangelo's, Giotto's, Botticelli's attempts to depict these divine beings has given me great pleasure, though if the truth be told none of these great artists ever evoked the awe and love conjured up by the wide-winged angels of the prints in my nursery, to whom we children lent such serene and protective powers. Here they were, as they had been so long ago, beautiful and smiling. It was wonderful to see. These men may have been naturally handsome, but that morning they were far more than that, creatures of an essence that wasn't of our world: their expressions of happy confidence as they got ready to ascend into their element, the sky, left me inspired, awed and earthbound.

The second account was by 'Mouse' Cleaver, one of the escort pilots:

PARIS, Saturday, June 1st, 1940

We were on ordinary readiness at Tangmere, and got a signal to go to Warmwell and pick up an escort job, which duly appeared, and we found ourselves a while later in Paris, when we discovered that it was Churchill. We were later told he was staying the night and we could go to town, takeoff next day 8 o'c. Archie managed to borrow quite a lot of money from a pal in the Embassy, and we set off for Lust and Laughter.

The next day there assembled at Villacoublay just about as hungover a crew of dirty, smelly, unshaven, unwashed fighter pilots as I doubt has ever been seen. Willie [Rhodes-Moorehouse] if I remember right

> was being sick behind his aeroplane, when the Great Man arrived and expressed a desire to meet the escort. We must have appeared vaguely human at least, as he seemed to accept our appearance without comment, and we took off for England.

Churchill was desperately seeking help from anywhere, if only to save time, but this last attempt to persuade the French to hold out hadn't succeeded. According to Billy Clyde, as they prepared for take-off he said how sorry he was he had failed them, then seeing their red eyes he added, 'I see at least *you* had some fun.'

The British government had promised before the war to send to France four fighter squadrons on the outbreak of hostilities. This was promptly done in September. In the months that followed, these hopelessly outnumbered units were supplemented until, by May of 1940, there were twelve squadrons operating from French bases. Dowding, as Fighter Command chief (and another ex-champion skier), was vociferously opposed to the drainage of fighter units to the Continent. He was convinced that the decisive battle would have to be fought over Britain and that the continued flow of resources to France was an utter waste. So badly did things go, nevertheless, that on 16 May he was unable to prevent eight more half-squadrons being added to the now pointless air defence of France.

Those pilots who crashed in France and had to leave their aircraft and return by land and sea were appalled by their impression of the Allied ground forces. 'Refugees blocked main roads and there was a sense of complete confusion,' wrote Huseph Riddle. 'Soldiers without rifles roaming aimlessly about the countryside, but all making their way westwards. Telephone communications were almost nil.' The pitiful demoralisation of the Allied armies was visible even from the air. Pilots reported the French soldiers would run panic-stricken from low-flying aircraft of either side. Only the Guards showed better fibre, for they stood firmly in the open firing their rifles – though indiscriminately at German and British aircraft alike.

When Churchill and his colleagues met with French premier Reynaud and General Weygand, the French view was that they were beaten unless more Hurricane squadrons were sent to repel the advancing German army. There was no air expert present to point out that Hurricane machine guns would be like pea shooters against a panzer, or to consider where the squadrons would be stationed, who would mount guard over them, or how they would be refuelled and rearmed. It was a naïve idea, but Churchill had nevertheless agreed to send help. Dowding objected, strongly but futilely. Men and machines sent to this lost cause would suffer irreplaceable losses while he still didn't have enough for home defence. He sent over half-squadrons, to be combined there, leaving part of every squadron intact at its home base.

One of the half squadrons sent over was 'A' Flight, led by Hope, which formed a composite unit with No. 3 Squadron at Merville. The other was 'B' Flight at St Valery, under Max Aiken. Their task was to protect the demoralised British troops retreating in disarray to the ports of Dunkirk and St Valery respectively.

Not a single 601 aircraft had been hit since the war began. Two days at Merville changed that. After an abortive patrol over Brussels on the first day, from which Guy Branch, having lost formation in the gathering dusk and force-landed, returned to camp drunk and inarticulate with an affluent Frenchman who had rescued him and then opened the wine cellar, the first German aircraft were engaged. The same day Hope followed the leader of No. 3 Squadron in an attack on Dorniers. To his surprise the section leader broke upward – a normally suicidal manoeuvre as it presents the target's rear gunner with a large, no-deflection target – and in the frenzy of the moment Hope presumed this to be some new technique known only to the regulars and followed suit. His aircraft was struck by return fire from one of the Dorniers and he crash-landed in an open field, returning to Merville in a motor cycle purloined from a neglected military dump at Amiens.

Merville had unique significance for Rhodes-Moorehouse. Twenty-five years earlier, his father, Second Lieutenant W. B. Rhodes-Moorehouse, took off to attack the Courtrai railway, pressed his attack home despite wounds in the stomach, thigh and hand, and died on landing at Merville to become the first RFC man to earn the Victoria Cross.

The Merville flight ran into a formation of He 111s returning to Germany from a raid on France. The Hurricanes, badly positioned for attack and frustrated by patches of cumulus that intermittently obscured the bombers, barely damaged the Heinkels but were damaged by return fire. According to Loel Guinness's combat report, 'My aircraft sustained the following damage: a bullet through the windscreen expended itself on the armour plating behind my head and went out through the hood, my hydraulics were shot away, the rudder badly holed, the fuselage struts shot through, the propeller boss shot through and there were various other bullets in the wings. The aircraft had to be left at Merville, as the engineering officer didn't consider it fit to fly.' Each aircraft was so badly hit that it was 'not considered fit to fly' on landing.

The Allied position was hopeless. Churchill, ever a Francophile, finally acknowledged that France was defeated (and would soon order the destruction of the French fleet in Oran to prevent it passing into German hands, and as an unmistakable signal of Britain's determination to fight on). When at last 'A' Flight was ordered back to England, Hope, charged with getting the aircraft back but fearing being swallowed up as a permanent part of No. 3 Squadron to which it was presently attached, on short notice and his own authority took the Hurricanes back to Tangmere. Cleaver, in charge of the ground party, arrived at Boulogne with the NCOs, airmen and equipment, to find no reception committee or organisation, and no boat until the following day. He billeted the entire contingent in an evacuated brothel in the centre of town, where they slept well as thousands of retreating soldiers huddled against their kitbags on the pavements outside.

There was only one problem with Hope's hurried evacuation to Tangmere. Huseph Riddle was still in the air, and when he returned from combat found Merville deserted. His Hurricane was out of fuel. An army lorry rushed over the airfield to him and its driver suggested he climb aboard at once as German forces were very close. Huseph tried to remember how to activate the self-destruct device in the Hurricane, which would render it useless to German hands. He couldn't. He turned the handles and switches on the device and stood clear. Nothing happened. By now the soldiers were threatening to leave without him. In desperation he drew his service pistol and fired all its rounds into the aircraft. No explosion or fire materialised so he shrugged and climbed aboard the lorry.

'Where is the RAF?'

The question born on the dusty roads to the beaches of Dunkirk on which the retreating British army was converging, spread from the Channel ports with the dispersing evacuees until it was echoed by Fleet Street and even politicians who should have known better. War in the air was widely misunderstood. Pilots returning wearily to their bunks from a day's engagements over France were able to buy newspapers from the old man on the camp on which the catch phrase was flaunted. It was infuriating and puzzling, because 601 was certainly not the only squadron that had suffered losses in the dangerous skies over St Valery and Dunkirk.

Besides those who were forced down in France to find themselves queuing with hostile soldiers on the beaches, there were those who never returned. Air fighting took place out of the retreating soldiers' sight but was often ferocious. On 13 May Demetriadi, Cuthbert, and Young were detached at a moment's notice to join No. 501 Squadron in France. Cuthbert was killed within an hour of arriving in France, and Lee-Steere on the 27th while covering the BEF withdrawal to Dunkirk. Charles Augustus Lee-Steere came from the cream of Australian society, the son of early settlers of the Swan River Colony.

Also lost over the same area within a few days of each other were Squadron Leaders Roger Bushell and Mike Peacock, each at the time leading his new squadron. Peacock's wry account of his penultimate flight was characteristic. Falling upon seven He 111s in Vee formation over Lille, instead of attacking the wing men as is usual he flew directly up the centre of the Vee and shot down the leader 'so that', he later explained, 'all the others, who had been too busy formating to look at their maps, would then get lost'. Peacock's aircraft was itself hit and he baled out safely. Next day, leading his squadron in an attack on ground vehicles, he blew up a petrol storage tank, taking himself and his Spitfire with it.

Though Mike Peacock's war effort was at an end, Bushell's was in a sense about to begin as one of the most troublesome prisoners the Germans ever took. Shot down by an Me 110 over Boulogne, Bushell crash-landed in a field. On clambering from the plane, which had broken in two, and thinking he was in friendly territory, he hailed a distant motorcyclist. This turned out to be a German dispatch rider who took him prisoner at pistol point.

The front line was moving so quickly that picking friendly territory for a forced landing wasn't the easiest thing a pilot could do. Hope was lucky to reach a stretch of beach that wasn't in enemy hands when he was forced down by the fire of an Me 109. He was surprised from behind while attacking a formation of Dorniers and, 'like a damned fool, pulled the wrong way!' A twenty millimetre shell burrowed into each wing root, a foot either side of the cockpit, and smaller bullets hit the engine cooling system. As glycol (engine coolant) streamed from the engine and its temperature rose sharply, Hope knew it was only a matter of minutes before his engine would seize up, and he headed for the flat beaches away from Calais. Having put his aircraft down smoothly and set fire to it, he was taken by French civilians to a British brigade headquarters south of Dunkirk where he was the object of scant attention as the air raid sirens had just sounded. A colonel took him by the elbow and led him towards the shelter with a paternal, 'Come on, old boy, Jerry's at it again.'

Hope shielded his eyes to squint at the fighters overhead. As they banked the late afternoon sun caught the distinctive half white, half black British recognition markings on their undersurfaces.

'Those', he replied, 'are Hurricanes'.

The colonel looked at Hope quizzically, suspecting a leg-pull. A major nearby, having overheard, asked, 'How the devil can you tell that?'

It was Hope's turn to be surprised. 'By the black and white colouring underneath. They're painted like that so you'll know they're ours. I happen to know they couldn't be Spitfires, so they must be Hurricanes.'

The brigade staff assumed thoughtful expressions and, after a pause, one of them said, 'But we've seen lots of those.'

'Yes, of course,' replied Hope, 'because you've seen lots of Spitfires and Hurricanes.'

At that moment another officer appeared, his gas mask slung over his shoulder. Seeing a man in aviator's garb, he threw a meaningful glance at the sky and then asked, with a frosty smile, 'Hello! And where is the RAF today?'

There was a gross misunderstanding of air operations. The army failed to understand that fighter protection occurred thousands of feet over their heads, out of sight and often out of hearing. The scream of Stuka sirens and explosion of bombs seemed to cry out for British fighters, close by and clearly marked by roundels, somehow protecting the soldiers upwards in the manner of anti-aircraft fire. Such ignorance was one downside of an independent air force, and greater efforts at force coordination should have been made earlier. The ground forces also lacked aircraft identification training.

Even so, as many as two-thirds of the British army was evacuated from Dunkirk, far exceeding Churchill's expectations, though minus their equipment. It was a massive defeat, but suggestions that the RAF hadn't played its part were little short of monstrous. Historians, while acknowledging the contributions of the French army, the Royal Navy, and the many small civilian boats, agree that without Fighter Command all would have been for nothing and the evacuation would have been a disaster. The fighters had operated from dawn to dusk for all nine days, and on only two of these did the Stukas cause any serious harm to the troops. Churchill praised the performance of the air force in the midst of defeat.

General Weygand was now charged with the thankless task of defending France. He strengthened his forces along the Weygand Line in the north-east corner of France but the Belgian army had capitulated, and it wouldn't be long before the Germans crushed these defences and marched on Paris.

Dowding continued to fight his own battle with the War Cabinet. An uninspiring figure, he had been given control of Fighter Command when it was the cad of the litter because only bombers could win wars. His philosophy was simple. He wasn't trying to win the war, merely to avoid losing it. The nation was in dire peril of invasion, and only air superiority could prevent it, which meant fighters and more fighters. He opposed even the temporary transfer of squadrons to France, which he considered a lost cause, and insisted at serious risk to his career on a minimum of fifty squadrons in England. A socially awkward and intensely focused man, understandably nicknamed 'Stuffy', his obduracy didn't make friends and frustrated Churchill. He only got his way after delivering a forceful memorandum to the War Cabinet on 15 May. Often quoted in full, it is the last of ten career-risking paragraphs that make his case with many 'ifs' and a chilling conclusion.

> I believe that, if an adequate fighter force is kept in this country, if the fleet remains in being, and if Home Forces are suitably organised to resist invasion, we should be able to carry on the war single-handed for some time, if not indefinitely. But, if the Home Defence force is drained away in desperate attempts to remedy the situation in France, defeat in France will involve the final, complete and irremediable defeat of this country.

Final, complete and irremediable defeat of this country were words the War Cabinet neither wanted to read nor dared ignore. There would be no further transfers of fighters to France. One hundred and ninety-five Hurricanes had been sacrificed to no purpose.

Chapter 6

The Battle of Britain

Our task is to hold secure for all time the superiority which Germany has obtained over all countries in the air.

Herman Goering, address on 1 March 1939

It seems quite clear that no invasion on a scale beyond the capacity of our land forces to crush speedily is likely to take place from the air until our air force has been definitely overpowered.

Winston Churchill – address on 18 June 1940

It seemed it should be easy. From the coast of newly occupied France, the Luftwaffe's commanders could see the thin white line of cliffs on England's southern coast a mere twenty miles away. Their thoughts were not hard to read, and not that absurd: they could crush England now. Or perhaps later, after they had dealt with Russia. Hitler was confident that Britain would be sensible and come to terms. His options were wide open. Everything he had sent his mighty Luftwaffe and Wehrmacht against had folded, one after the other: the Rhineland, Austria, Czechoslovakia, Poland, Denmark, France, Belgium, Luxembourg, the Netherlands. The might of Germany's armed forces had with little difficulty thrown Britain ignominiously off the Continent and now threatened her in all directions from air bases on the surrounding occupied territory. After listening to the encouragement of his generals Hitler allowed planning to proceed for the invasion of Britain, which would call at first for the elimination of Britain's air defences. A simple matter. 'It will take', reported General Stapf to General Halder, Luftwaffe Chief of Staff on 11 July, 'between a fortnight and a month to smash the enemy air force.'

Many others thought Britain had had it too, including Lord Halifax, the near-prime minister; Prince Edward, the ex-king; Charles Lindberg, the erstwhile popular hero, and Joseph Kennedy, the despised United States

Ambassador in London, soon to be replaced by Roosevelt because of his defeatism. Winston Churchill was an exception, and no matter how many times it has been said, he was the single most important individual standing between democracy and Hitler's evil intentions.

There would seem to be little to add to the mountain of literature and legend on the Battle of Britain. But much depends on where you stand. A citizen of London remembers mostly the Blitz, the homes destroyed, lives lost, and the nightly dread. To 601 that's when it ended. Their battle consisted of four punishing weeks of fatigue and horrendous blood-letting from 8 August, when the enemy's attacks on convoys and the south coast ports began, to 7 September, when its switch to the night raids pushed aside any chance of an invasion, which rough Channel conditions would by then render impossible.

Now twenty-six years old, and most of that time a bomber unit, 601 together with her sister squadrons waited for what was to come. Nobody quite knew what this would be, but as the British public girded itself for something terrible and drew on native resources of calm strength, the fighter squadrons in their forward stations along the coast sensed that, in their skirmishes with the enemy, they were being prepared for a battle of unprecedented dimension and ferocity.

The Prime Minister wasn't alone in appreciating the gravity of the situation, only in the clarity with which he defined it:

> What General Weygand called the Battle of France is over. I expect that the Battle of Britain is about to begin. The whole fury and might of the enemy must very soon be turned on us. Hitler knows that he will have to break us in this island or lose the war … Let us therefore brace ourselves to our duties, and so bear ourselves that, if the British Empire and Commonwealth last for a thousand years, men will say, 'This was their finest Hour'.

It is true that words can't win wars, but they can prevent a war from being lost. Like almost everybody else in the nation the pilots of No. 601 Squadron listened to Churchill's speeches expressing their own thoughts. They knew there would be momentous consequences for them.

There were appeasers in the British government who believed, quite rationally, that Britain's plight was so dire that terms should be reached with Hitler that would save the country from the ravages of bombing. For Churchill, saving the cathedrals and beautiful countryside was less important than saving the country's freedom. On 28 May 1940 he made a speech to the Coalition cabinet that galvanised it into united defiance and a determination, somehow, some day, to win.

As Goering and Churchill knew, the Luftwaffe had to gain air superiority. Without this the invasion barges would make easy targets for the RAF and the Royal Navy, while without barges the German army could not invade. Air superiority meant the destruction of Britain's air defence system: its radar stations, airfields, and fighters with mass attacks of bombers and escorting fighters. Dowding was determined to preserve his fighter force and sought to avoid its tangling with German fighters, which would be playing the Luftwaffe's game. Fighter Command's resources were strained, but less so than the Germans thought. This was how Dowding wanted it: for the Germans to believe he had a weaker hand than he did so as not to invite overwhelming attacks. He sent up units of only three or six, deliberately holding 12 and 13 Groups further north in reserve. While this guaranteed that No. 11 Group's pilots would always be outnumbered, its commander, Keith Park, fully supported the strategy. At the same time, fighter production was being boosted by the energy and organisational skills of Lord Beaverbrook, Minister of Aircraft Production and Max Aitken's father. Beaverbrook put fighter production above all else, to the anger of the 'bomber barons', and not only because like Dowding he had a son in Fighter Command. He understood Dowding was playing for time because by October the rough Channel waters would rule out an invasion.

The danger was grave. Von Rundstedt's army stood at the coastline of defeated France, and Goering proceeded to deploy his Luftwaffe on a wide arc of French, Belgian and Dutch airfields facing Britain from the east and the south – over 3,500 aircraft against 500 British fighters – in readiness for the invasion, Operation *Sealion*, which called initially for the elimination of air resistance. The evacuation of a third of a million British and Allied troops from the continent of Europe without their equipment, the fall of France and the entry of Italy on the 10 June to the European conflict, were pregnant enough with foreboding. But now the world's largest air force was within half an hour's flight of the world's largest target, and Churchill was saying we must win the war.

Hitler was sure that Churchill was bluffing, and he allowed a few weeks to pass for the British government to replace him and put out feelers for an armistice that, not being negotiated under violent attack, would preserve its honour. None came, for the Cabinet's decision to fight on was irrevocable.

The pause, during which only 5 to 10 per cent of Luftwaffe strength was used in experimental flights and convoy raids in the English Channel, though tiring enough to the fighter pilots of No. 11 Group, afforded Beaverbrook's ministry the opportunity to build up fighter resources.

Whenever a pilot was lost the strength of his squadron would be measurably sapped and his replacement became a matter of urgency. Among the Legion's replacements were an Australian and an American. The Australian was Flying Officer Clive Howard Mayers, a fair-haired citizen of Melbourne who would be ruffled when somebody mimicked his accent. 'It's not Austr-y-lia,' he would protest, 'it's Austr-y-lia'.

Although the American government discouraged its citizens from joining British armed forces, seven did so and, including those already in England, eleven American pilots were to fly with Fighter Command during the Battle of Britain and two with 601. One who had already joined 601was Carl Raymond Davis, born in the Transvaal to American parents and already an established Legionnaire. The other was Bushell's friend Billy Fiske.

Fiske had studied at Cambridge in the late 'twenties and early 'thirties and used to drive his golfing friends the twenty-four miles to Mildenhall in seventeen minutes in his 4½-litre open Bentley, changing up at eighty and said never to vary more than two or three inches in his road track. It was no ordinary Bentley, if that's possible, but a blower (supercharged), rebuilt to Bentley team standards, with a long bonnet and coloured British Racing Green. After Cambridge, together with a partner he bought an abandoned mining building in Aspen, Colorado, and converted it into a ski lodge similar to the ones he had seen in Europe. This was the beginning of skiing at Aspen. Friendships forged on the ski slopes of Europe with Bushell and Clyde had prompted him to make the big jump. After disembarking from the *Aquitania* Fiske applied to join the RAF. He was accepted, not surprisingly given its urgent need for pilots and his evident skills, and passed comfortably through all stages of his basic and operational flight training at RAF Yatesbury and Brize Norton. When commissioned he expressed a preference for fighters and predictably for No. 601 Squadron.

In his book *The Few* Alex Kershaw writes that the British government saw Fiske as 'an extraordinary asset. In their eyes, his celebrity was now a far more potent weapon than any plane he might fly' and he was called to a meeting with Air Minister Sir Archibald Sinclair, who asked if he would be willing to embark on a propaganda tour of the States. 'It seemed that Fiske was in a double bind,' Kershaw writes. 'If he refused to do what Sinclair wanted, his career in the RAF would surely be doomed and he would be barred from joining his friends in 601. But if he accepted and returned to America, he would then be arrested for breaking neutrality laws.' Afraid of either offending the RAF and being turned down for 601, or of returning to the US and being arrested, he is said to have stalled, arguing that he would have more propaganda value after he had shot down some Heinkels, and that Sinclair saw the sense in this.

This can hardly be true. The US Neutrality Act of 1939 imposed an embargo on trading in arms and war materials but did not ban citizens from participating in foreign wars as long as they were not belligerents of the United States or threaten to arrest them if they did. Instead, it issued a caution that if citizens entered a war zone they did not expect the protection of the United States. The retired USAF pilot Claire Chennault had been advising Chinese loyalists since 1937 and was renowned for organising the American 'Flying Tigers' squadron in China, while American pilots fought on both sides during the Spanish Civil War. Moreover, it seems improbable that the British government, after training Fiske to fly and qualify for RAF wings at a time when Britain was desperately short of fighter pilots, would want to throw this away on a speculative propaganda mission.

Pilot Officer Billy Fiske arrived at Tangmere, with no pretensions and no illusions, on 12 July 1940. He moved into a manor house near the airfield with his beautiful English wife Rose Bingham, who had followed him over. Described by the press as 'the prettiest titled lady in England', she was a former wife of the Earl of Warwick and a Hollywood actress whom Billy had met at St Moritz. Some writers have supposed that the CO and his flight commanders on 601 had some initial misgivings about this untried adventurer, this jumped-up Yank with little or no flying experience who was going to show them how it was done. Kershaw writes that Jack Riddle said, 'We were pretty disgusted when we heard this pretty kid was coming over from America to tell us how to fly Hurricanes and how to shoot down Messerschmitts.'

This is hard to believe. It would have been out of character for Jack Riddle to speak this way of anyone; he wasn't a xenophobe, and those who knew him never heard him express anything but admiration for Billy Fiske. Besides, it makes little sense. Fiske was no swaggering Yank but a society figure ('celebrity' in today's parlance) and a polished man of the world in the mould of his close friends Bushell and Clyde. He had won gold medals in Europe, attended Cambridge, was an obvious Anglophile, and had an aristocratic English wife. If being famous, rich, and accomplished in motor racing and sports still counted for anything he would have been judged eminently qualified to join the Legion. The leading members of 601, many of whom had some American blood in their veins, had urged him to do so. His reputation was already high and he was welcomed with open arms.

Fiske was quickly liked as a man and admired as an aggressive and fearless fighter pilot. As his flight commander Hope would later say, 'In all my flying experience I have never come across a pilot with such completely natural flying ability and quick reactions. He made his aircraft become part of him and had the potential of an ace.' Fiske quickly absorbed squadron life while Rose mingled happily with the other wives. Like the others, he enjoyed a drink or two at The Ship. He was often joined there by Jack Riddle, and the two became close friends. Sometimes their wives would join them, and Riddle's wife once asked how it was that Fiske always seemed to get there first. Riddle simply smiled and pointed out of the window at Billy's 4½- litre Bentley in the parking lot.

Each pilot flew two days with one off, a lorry taking him to dispersal on the airfield before dawn and bringing him back after a day's flying and air fighting – unless he was on night readiness – for a pint in the mess and a badly needed night's rest. During the day, while waiting for the scramble order, the pilots lay in the grass, or sat at a trestle table drinking tea from real teapots and real cups and saucers, or remained in the crew room reading or playing darts or shove-penny. The radio would usually be playing and they would often listen to the news and always to Winston Churchill's speeches. They wore their Mae Wests at all times.

The Hurricanes were parked nose-into-wind, their engines run up periodically to keep them warm between sorties. The different categories of preparedness in themselves introduced an element of strain, an invisible tether of varying radius broken only by the end of the day or the bustle of a 'scramble'. Often there was no time to remove the cumbersome flying clothes or go very far, although one would probably spend hours inactively, and one of the fighter pilot's greatest anxieties was the possibility of being scrambled while on the lavatory. Archie Hope took the opportunity of a bath one late afternoon after satisfying himself with a phone call to Station Operations that nothing was brewing, and had a sponge in his hand when he was scrambled, making it after pulling on his clothes without pausing to dry himself in just over five minutes.

As the great battle drew to a close there was a mood of calm expectancy. It was anyone's guess whether these happy and privileged young amateur flyers were ready for brutal, sustained war. The betting had been that they would, just as Territorials always had. The Germans didn't think so; they doubted the rich amateurs' piloting skills and warrior spirit. They also, with more justification, doubted Fighter Command's tactics.

The Luftwaffe had more pilots than the RAF, with better training – about thirteen months' with 150 to 200 flying hours – whereas the RAF couldn't wait to rush green pilots into service. The RAF 'Vic' or Vee-shaped formation initially designed for tight bombing patterns was continued for fighters for two reasons: first, to bring concentrated fire on enemy bombers, and second, to make newly qualified pilots follow their leader without having to think. Initiative was not only inhibited but at first forbidden; attacks were to follow numbered, set-piece procedures. Close formation, though a fine exercise in concentration and control, presented escorting fighters with steady targets that were too closely focused on each other to look around. It took serious losses to unlearn these theoretical tactics. The Germans called the Vics '*Idiotenreiche*'. In reality the tidy little Vic formations quickly broke up into dogfights, but the enemy's escorting fighters took full advantage of the RAF's momentary indecision at a time when seconds counted. The RAF's pilots would learn, some sooner than others, that precision flying was of little use and even potentially suicidal in dogfights. On the other hand coarse handling avoided presenting a steady or predictable target. As the RFC fighter Ace "Mick" Mannock had said in the first world war, 'Good flying never killed a hun'. Auxiliaries and regulars alike were thrust into the fray without this awareness, against the world's most experienced and confident fighter pilots. The Luftwafe used a '*schwarm*' of two loose pairs developed during the Spanish Civil War, in which the number two of each pair kept a lookout while number one went for the kill, and each pair was able to pull away and act independently as needed. Though this formation was highly effective, Luftwaffe number twos grumbled that they had to use more fuel following, were more likely to be killed, and never got a chance to score while their number ones went on to high scores, medals, and glory.

The Luftwaffe faced other problems, the biggest of which, aside from the misfortune of being led by the sycophantic and drug-addled Goering, was that without long-range droppable fuel tanks the endurance of Me 109 fighters was only about an hour, leaving little time for combat. This short endurance coupled with two long flights over water caused the pilots an anxiety they called *Kanalkrankheit* or 'Channel sickness'. Given this added stress nobody could deny the courage and skill of the Luftwaffe aircrews.

The first phase of the German air assault began with three raids in the morning, midday and late afternoon of 8 August, on Channel convoys and south coast ports. The attacks were intended to draw the RAF out so they could be dispatched into the sea by overwhelming numbers of escorting fighters, their pilots lost forever. The Germans flew low, where radar couldn't detect them, luring out the RAF fighters. Dowding was worried that brawls would erupt; he had no interest in destroying German fighters but in preserving his own to bleed the enemy bomber force to death. Although this was simply an intensification of the preceding raids, 8 August was for a while considered by the defenders to mark the beginning of the Battle of Britain, so severe was the fighting. All day long the raids were countered by five Hurricanes and two Spitfires of 11 Group, the group which by virtue of its geography was to bear the brunt of the coming onslaught.

The lovely, rolling countryside of southern Kent around Folkestone, Dover and Lympne, which had been the scene of the Legion's most carefree days during the 1930s, at Botolph's and the Cinque Ports Club, came to be known by the RAF as 'Hell's corner', an unusual epithet for such embracers of understatement. At this narrowest part of the Channel the Luftwaffe would have the least 'Channel sickness' and thus the longest time available for combat; it was where, if you were thirsting for action, you should be. You couldn't miss it –barrage balloons, bomb and shell bursts, condensation trails, parachutes, streams of smoke, splashes in the water, and aircraft burning in the air and on land.

Along this stretch of land and water over which the battle raged that day, twenty-eight German and twenty British aircraft came down for the last time. Among the thirteen RAF pilots killed was twenty-seven-year old Flying Officer Guy Branch, the quiet, slightly built fellow who had illuminated the bathroom of his London flat with candles stuck in bottles, and who had won the Legion's first decoration, the Empire Gallantry Medal, for saving Aidan Crawley's life.

Larger formations of enemy aircraft – Dorniers, Heinkels, Messerschmitts, Stukas – gave the squadron a foretaste of the mass raids to come. It was no longer necessary, as at Merville, for the fighters to queue up to attack compact formations of the enemy. Hope led a formation of five from which only he returned to Tangmere, the other four being forced down though with no loss of pilots. In another engagement Peter Robinson was shot in the leg by an Me 109, forced down and strafed on the ground. One Hurricane was fired on by another Me 109, escaping into a cloud with smashed instruments, ailerons and flaps.

Although the real test was yet to come, the pilots were getting tired. Sometimes dizzy from lack of sleep, limbs aching from long hours strapped in tiny cockpits, they took off, flew, fought and landed almost automatically and hoped that the crews of *Luftflotten* 2 and 3 were equally weary. Life was full. The AOC 11 Group, AVM Park, visited Tangmere, and when he had gone Aitken wrote in his log book: 'Flew five hours before lunch. AOC asked, "Are you tired?"'

Five days after Villacoublay and three months after the Hurricanes arrived, Loel Guinness, at the fighter pilot's old age of thirty-four, was promoted to a controller's job and succeeded by Max Aitken. Within three weeks Aitken scored a notable victory for which he was awarded the DFC. Night fighting, as such, didn't exist. Little could yet be done about the occasional raider that crossed the coast in darkness. Nevertheless, on Churchill's orders in response to public and political pressure and over Dowding's realistic objections that the necessary airborne radar and Beaufighters to carry them were not ready and it would be throwing away aircraft and pilots, readiness was maintained in rotation by ordinary day-fighter squadrons, and on 26 June Aitken and Tom Hubbard scrambled at eleven o'clock to intercept a plot near the patrol line. There was good cloud cover in patches but a strong moon, and by chance a battery of searchlights picked up one He 111 in a formation of three. Before the bomber could escape the beam, Aitken slipped behind it and with a long burst between its illuminated exhausts sent it into the English Channel. There it floated, ringed with foam, the crosses on its wings showing clearly in the light of Aitken's flare. This was a rare achievement, but it caused Aitken to leave 601 because within a month he found himself given command of No. 68 (Night Fighter) Squadron.

The night fighter squadrons were Churchill's response to the night intrusions because he felt compelled to do something, though little really could be done. Dowding argued in vain for airborne radar and predicted a high accident rate to no avail. Max Aitken described this period as for him, 'perhaps the most potentially dangerous in the war. [Barrage] balloons were up everywhere, the countryside was completely blacked out and our aircraft were crashing all over the place. Giving dual instruction to a frightened pupil was a nightmare. I had to change seats with my pupil – no pleasant thing on a dark night in a confined cabin [in a Blenheim] – six or seven times.'

So many air battles were fought over the Channel that survival in the sea and air-sea rescue became vital concerns. In this respect the Germans were well ahead. The squadron remembered that it had been packed off to Borkum without rubber dinghies or life vests; they now had yellow Mae Wests that would inflate automatically but still no dinghies and insufficient high-speed launches, with the result that many RAF pilots died of hypothermia bobbing up and down within sight of the Sussex coast and of their comrades overhead. The Luftwaffe had made good provision for rescue at sea. Pilots had life jackets, dinghies, dye markers, a greater number of rescue launches

– and seaplanes. These rescue seaplanes, bearing red crosses, capable of locating a downed flyer and landing close to him, provided a fast rescue service. Dowding worried that these seaplanes were useful for reconnaissance over the convoys and especially that recovered aircrew would return to the Luftwaffe and be free to fight again. He found the air-sea rescue disparity unacceptable and on 14 July gave orders for the German seaplanes to be fired on, whether airborne or on the water. Few RAF pilots actually carried out this order, though some did. Their advantage pruned, the Luftwaffe retaliated with the practice of firing at RAF pilots in their parachutes and in the water, which was at first occasional but became more frequent in later stages of the war. The difference between pilots parachuting into the sea or over England was so crucial to the Fighter Command's resources that Dowding discouraged air fighting over the Channel as well as dogfights with enemy fighters.

Three 601 aircraft spotted one of the He 59 twin-engined, twin-float seaplanes flying suspiciously near a British convoy and about a hundred feet above the water. It was all white, with large red crosses and the civil registration markings DA-KAR. Flying Officer Tom Hubbard, somewhat non-plussed, called the controller and was told to shoot it down if it was hostile. The Hurricanes, by making menacing passes without firing, signalled the seaplane to head for England, and this it did for a while, only to turn south again. A few warning bursts having no apparent psychological effect, they shot it down. Before the Heinkel hit the water its crew of four jumped out but they were too low for their parachutes to open.

Good controllers with fighter experience were much in demand. Wing Commander Eddie Ward, the twin brother of Geordie Ward of Sassoon's days and a friend of the Legion, was a controller at Southern Sector. Controllers had great authority and the rank to enforce it. He directed a formation led by Hope to orbit over Bosham. Ward reported that 'bandits' were approaching the Hurricanes but the pilots could see none, causing general consternation. When he confirmed that the 'bandits' were directly over Bosham, faith in the radar system almost collapsed. But Hope, suspecting the human element, threw security to the winds. 'We're not only over Bosham, Eddie,' he called, 'we're directly over The Ship – is that where you want us?'

'Roger!' affirmed Ward, 'Over the ships.'

Noting the plural and embarrassed at his mistake, Hope took his formation to the convoy of merchant vessels south of Selsey Bill code-named 'Bosom'.

No. 601 Squadron continued to provide convoy escorts over the Channel. Somtimes the action wasn't intense, or there was often no action at all.

Starting at dawn the Hurricanes would orbit within sight of the convoy. After a couple of hours, with fuel running low, they would look forward to returning for breakfast in the mess. They began to call the base individually on the R/T to place their breakfast orders, such as bacon, egg (always only one egg), beans, and tomatoes. This practice annoyed Tangmere's station commander, who considered it a misuse of radio communication and ruled that it must stop. It did stop, but then tentatively resumed after a few days, with no consequences. Why, nobody knew, but the pilots liked to believe the rumour that Churchill had approved of the practice because it suggested to any listening Germans that food rationing wasn't a concern.

The British people were told to keep calm and carry on. As demonstrated by the insouciance of the Richmond Golf Club's rules, this advice may have been unnecessary and offered some promise that there would always be an England:

RICHMOND GOLF CLUB TEMPORARY RULES, 1940

1. Players are asked to collect Bomb and Shrapnel to save these causing damage to the Mowing Machines.
2. In Competitions, during gunfire or while bombs are falling, players may take cover without penalty for ceasing play.
3. The positions of known delayed action bombs are marked by red flags at a reasonably, but not guaranteed, safe distance therefrom.
4. Shrapnel and/or bomb splinters on the fairways, or in Bunkers within a club's length of a ball, may be moved without a penalty, and no penalty shall be incurred if a ball is thereby caused to move accidentally.
5. A ball moved by enemy action may be replaced, or if lost or destroyed, a ball may be dropped not nearer the hole without penalty.
6. A player whose stroke is affected by the simultaneous explosion of a bomb may play another ball from the same place. Penalty one stroke.

With convoys in the Channel under attack and the mainland still untouched, a sense of unreality, even theatre, pervaded as the war was fought in full view of British spectators watching from the southern coast. On 14 July 1940 the BBC's Charles Gardner, a well known sports commentator, together with his crew set up sound recording equipment and a tripod with field glasses on the South Downs on the rim of Hell's Corner overlooking the Channel. In a breathless race-track style he painted a vivid word picture of an air battle that could be seen heard only in fragments: Ju 87s dive-bombing the ships, the splashes and explosions, RAF fighters attacking the dive bombers, flames, stricken aircraft trailing smoke, parachutes blossoming, the sounds of distant machine guns and thundering anti-aircraft fire. Like many others misusing 'Spitfire' to mean any RAF fighter, Gardner gave listeners a mental picture of aerial warfare as he perceived it. It was immediately controversial:

> The Germans are dive-bombing a convoy out at sea; there are one, two, three, four, five, six, seven German dive bombers, Junkers eighty-sevens. There's one going down on its target now. Bomb! No! he missed the ships, it hasn't hit a single ship … there are about ten ships in the convoy, but he hasn't hit a single one and … There, you can hear our anti-aircraft going at them now. There are one, two, three, four, five, six – there are about ten German machines dive-bombing the British convoy, which is just out to sea in the Channel … Here they come. The Germans are coming in an absolute steep dive, and you can see their bombs actually leave the machines and come into the water. You can hear our guns going like anything now. I am looking round now. I can hear machine gunfire, but I can't see our Spitfires. They must be somewhere there. Oh! Here's one coming down … There's one going down in flames. Somebody's hit a German and he's coming down with a long streak – coming down completely out of control – a long streak of smoke – and now a man's baled out by parachute. The pilot's baled out by parachute. He's an eighty-seven, and he's going slap into the sea – and there he goes. SMASH! A terrific column of water and there was a eighty-seven. Only one man got out by parachute, so presumably there was only a crew of one in it.

As would the public, Gardner assumed that every pursuing aircraft was British and every aircraft destroyed was German. He was unaware that Park had only three aircraft engaged, being Red Section from No. 615 Auxiliary Squadron. The Ju 87 he described was in fact a Hurricane and the pilot baling out was Pilot Officer Michael Mudie, who was later picked up by the Navy but died the next day of severe wounds. Gardner continued:

> Now, then, oh, there's a terrific mix up over the Channel! It's impossible to tell which are our machines and which are Germans … There was one definitely down in this battle and there's a fight going on. There's a fight going on, and you can hear the little rattles of machine gun

bullets. Grump! That was a bomb, as you may imagine. Here comes one Spitfire. There's a little burst. There's another bomb dropping. Yes. It has dropped. It has missed the convoy. You know, they haven't hit the convoy in all this. The sky is absolutely patterned with bursts of anti-aircraft fire, and the sea is covered with smoke where the bombs have burst, but as far as I can see there isn't one single ship hit, and there is definitely one German machine down. Well, that was a really hot little engagement while it lasted. No damage done, except to the Germans, who lost one machine and the German pilot, who is still on the end of his parachute, though appreciably nearer the sea than he was. I can see no boat going out to pick him up, so he'll probably have a long swim ashore. Well, that was a very unsuccessful attack on the convoy, I must say … Oh, we have just hit a Messerschmitt. Oh, that was beautiful! He's coming right down. I think it was definitely that burst got him. Yes, he's come down. You hear those crowds? He's finished! Oh, he's coming down like a rocket now. An absolutely steep dive. Let us move round so we can watch him a bit more. Here he comes, down in a steep dive — the Messerschmitt. No, no, the pilot's not getting out of that one. He's being followed down. What, there are two more Messerschmitts up there? I think they are all right. No – that man's finished. He's going down from about ten thousand, oh, twenty thousand to about two thousand feet, and he's going straight down – he's not stopping. I think that's another German machine that's definitely put paid to. I don't think we shall actually see him crash, because he's going into a bank of cloud. He's smoking now. I can see smoke, although we can't count that a definite victory because I didn't see him crash. He's gone behind a hill. He looked certainly out of control … Now there's one coming right down on the tail of what I think is a Messerschmitt and I think it's a Spitfire behind him. Oh, darn! Now – hark at the machine guns going! Hark! one, two, three, four, five, six; now there's something coming right down on the tail of another. Here they come; yes, they are being chased home – and how they are being chased home! There are three Spitfires chasing three Messerschmitts now. Oh, boy! Look at them going! Oh, look how the Messerschmitts — Oh boy! That was really grand! There's a Spitfire behind the first two. He will get them. Oh, yes. Oh, boy! I've never seen anything so good as this. The RAF fighters have really got these boys taped. Our machine is catching up the Messerschmitt now. He's catching it up! He's got the legs of it, you know. Now right in the sights … Go on, George! You've got him! …

> Yes, they've got him down, too. I can't see. Yes, he's pulled away from him. Yes, I think that first Messerschmitt has been crashed on the coast of France all right.

The scene is easy to visualise, but much of it was speculative, even banal. He did as well as anybody might but his imagination, like his listeners', couldn't encompass the bedlam, the fear and pain, the intense physical and mental stress, and the obscenities in English and German which rocked the airwaves, the literal blood, sweat, and tears of Churchill's metaphorical promise. Some listeners complained that his tone was unduly flippant and his relish at the loss of human life unseemly. Others could see nothing wrong with it.

The civilian populations of the coastal towns had their own share of action as the Luftwaffe bombed the ports. Ann Attree (formerly Mrs Ann Davis) of the Cinque Ports Club, who in 1935 had bought the wreckage of Aitken's little Aeronca, wrote to the squadron after a severe bombing, 'It looks exactly as though 601 have had another guest night.'

Weather prevented the Germans from launching their main offensive as scheduled on 10 August. Although the 11th, when 601 helped at no cost to itself to destroy thirty-five enemy planes for thirty-two British, was a busy day, the Germans considered that the real battle began on the afternoon of 13 August. To 601 it seemed otherwise. They were now having to throw themselves without hesitation into such vast formations that they hardly knew where to start, and total disregard for such odds became commonplace. This spirit began to cost the squadron dear. Eight of its Hurricanes attacked a heavily escorted formation off the Dorset coast on 12 August, were themselves attacked from all sides, and paid for the destruction of twelve German aircraft with the lives of Flying Officers Demetriadi, Smithers, Gillan and Dickie. The abrupt loss of four pilots, hopelessly outnumbered in the kind of fighter-to-fighter melee Dowding had sought to avoid, meant 601 Squadron lost a fourth of its fighting force in one day. It was a brutal foretaste of aerial combat against superior numbers of better trained, better equipped, more experienced, and equally brave German aircrew.

On 13 August, *Adler Tag* or 'Eagle Day', the Germans launched the first of their major attacks, choosing as objectives airfields in the south and south-east and widening the front to test the strength of the fighter defence. They found the RAF able to fight over two counties simultaneously, and the raids were ineffectual. The squadron attacked, harried and split up the bombers in three fierce battles – in the early morning over Haslemere, at noon over

Portland Bill and in the later afternoon over Southampton. Not one pilot was lost, and 601 shared with No. 609 Squadron the highest honours of the day when each was credited with thirteen of the enemy. Although the Luftwaffe believed that four RAF aircraft had been destroyed for each of their own, the actual score was thirty-six German for twenty-two.

The noon battle was, in terms of numbers of aircraft to fall out of the sky in a short time, the fiercest 601 had ever known. The experience of Clive Mayers, described by him and Archie Hope in their official reports within minutes of returning to Tangmere, is an example. Hope led the scramble and his section dived in line astern into a tight pack of bombers at twenty-one thousand feet over Portland Bill. The Hurricanes were in turn attacked by the escort and a typical dogfight began. The air was soon bruised with brown anti-aircraft puffs and criss-crossed with condensation trails; pillars of smoke marked the departure of victims, blazing fragments fell from aircraft that had exploded in the air, and here and there parachutes blossomed. One of the parachutes belonged to Mayers. He had damaged an Me 110 and was watching it enter cloud at seven thousand feet when:

> My Hurricane was hit by what seemed a tornado. I felt pain in my right buttock and leg, felt the engine stop, heard hissing noises and smelt fumes. My first reaction was to pull back the stick and there was no response. The next thing I remember was falling through the air at high speed and feeling my helmet, flying boots and socks torn off. Lack of oxygen must have dulled my senses as the combat ended at 19,000 feet and my parachute opened just above the clouds at 7,000 feet. At about 5,000 feet between two layers of cloud an Me 110 fired at me while being chased closely by a Hurricane.

The pursuing Hurricane was Hope's, and as the fight subsided he called over the R/T for others with sufficient fuel to help him direct the rescue launches to the numerous pilots, both British and German, in the water. The launches were hampered in their search by a heavy swell and the absence of R/T. Without the few Hurricanes, which flew over the launches in the direction of the floating airmen, several pilots would have perished. 'I am convinced', reported Hope, 'that unless we had circled these pilots in the water at least three of them would not have been picked up. They were easy to see from the air as long as their parachutes were floating, i.e., half an hour at the most. It would have been helpful if we could have dropped some kind of marker.'

Mayers was among the first to be rescued:

> I landed in the water [according to his operational report] about three miles from Portland. Just before landing I saw an MTB [motor torpedo boat] about a mile away but the CO [of the MTB] told me later they didn't see me coming down although they saw a German parachutist not two hundred yards from me. After about twenty minutes I saw a Hurricane searching the bay and recognised it as belonging to my Flight Commander, Flight Lieutenant Hope. He waved to me and spent some considerable time in attracting the attention of the rescue launch … I am quite sure that if it hadn't been for Flight Lieutenant Hope the MTB wouldn't have found me.

Hope then directed the launch to the German pilot, lying face downward and apparently lifeless, who proved when fished from the sea to be very much alive. Mayers was towelling himself and sipping hot cocoa when there was a commotion on the deck. A seaman, with good intentions and a sharp knife, was about to cut the life jacket off the Luftwaffe pilot, but the German, thinking he was about to be murdered, put up an obstinate fight. He was allowed to keep his wet clothes on.

After some night activity, 14 August was quieter. The raids were shorter with only one mass raid over Kent, as the Germans conserved their energies for the following day, pivotal in the *Sealion* timetable. Billy Clyde and Johnny McGrath were shot down, but both were uninjured. Clyde fell into the Channel and was rescued after a prolonged wait in the cold water that permanently affected his health. They brought him back to shore, shivering and near death. He was removed from operations and, with a record of nine kills, awarded the DFC and sent to Washington with the rank of wing commander to be part of the Combined Chiefs of Staff. In the same battle 'Mouse' Cleaver, his eyes filled with Perspex fragments from the hood of his Hurricane from debris while attacking a Heinkel, managed to bale out and parachuted down at Lower Upham, near Southampton, from where he was taken to hospital. Now blind in his right eye and with limited vision in his left, he would never return to flying. He was awarded the DFC, and his eye wound would in due course lead to a profound medical innovation.

The Luftwaffe had been waiting for their big chance: concerted and highly organised attacks of maximum force upon the English mainland, which, since the occupation of Denmark and Norway, was now menaced from a wide crescent of Luftwaffe bases. If the RAF had been under strain during the previous weeks, they reasoned, its resistance should collapse before an all-out effort along a wider front than anything yet attempted. Everything was ready but the weather, and on the 15th that, too, was all set, and Goering's commanders studied with satisfaction the latest meteorological reports from night-flying weather planes. In the biggest aerial assault of all time the German spear point widened into a hammerhead as 1,790 bombers and fighters converged on England in an arc stretching from the Tyne to the Exe, aiming to efface the supposedly tenuous line of defence with five heavy and several diversionary raids. Supposing this area to be undefended with all RAF reserves pulled down south, they sent unescorted bombers from Norway. The fighting lasted eight hours along a hundred-mile line. Nos. 12 and 13 Groups were ready and the attackers' losses were heavy, for minimal damage done. Five attacks were repelled by No. 11 Group, beginning with 110 aircraft over Folkestone and Lympne at half-past eleven in the morning and continuing with raids of varying force, some of which reached the greater London area, until seven-thirty that evening. When in the afternoon seven formations comprising 300 raiders approached the shore of Hampshire and Dorset, 601 joined the heaviest fighter force the RAF had ever put up, eight squadrons of Spitfires and Hurricanes that intercepted the enemy over the Channel. The squadron attacked a heavy formation of Ju 88s from its starboard side, at twenty thousand feet some miles south of Spithead. From the first attack, three were shot into the sea and two made to jettison their bombs. The 'Millionaires' Squadron' destroyed five Ju 88s by midday.

Those aircraft that crossed the coast were harried and split into groups that were easier to deal with; their numbers further reduced as bombers were shot down or forced to land. Those that continued often missed their targets or were caught on their return by fighters that had refuelled and rearmed. Even when out of ammunition the Hurricanes and Spitfires showed themselves capable of effort. Hope made violent passes at a Ju 88 'to try and frighten him and to stop him going south', pleading over the R/T for someone to finish the matter with a five-second burst, when the bomber's crew unexpectedly chose to await the invasion in England and took to their parachutes. Billy Fiske, his bullets also gone, successfully manoeuvred a straggler into the Portsmouth balloon barrage.

The day had been an unquestionable triumph for Fighter Command. The defence had held, no widespread damage had been done to ground targets, and as many as seventy-six enemy aircraft had been destroyed for the loss of only thirty-four. Most of the German planes brought down contained a composite crew, and Goering immediately ordered after the results of the

15th that no aircraft operating over England was to contain more than one officer in its crew. 'Bombing England,' Churchill had warned, 'will be no child's play.'

Astonished and puzzled by the stiff resistance, the Luftwaffe resolved overnight to concentrate attacks upon the Royal Air Force. This was a sound stratagem, but coupled to it was a misjudged change of plan in the abandonment of attacks on radar stations, which Goering erroneously believed had been seriously damaged. The 1,720 aircraft that raided England on 16 August had airfields as their targets. No RAF base was closer to the attackers than Tangmere. The nature of the struggle for air superiority was now clear: the German strategy was to destroy the British fighter force; the British strategy was to destroy the German bomber force.

A *Staffel* of Ju 87s rose from its base in Normandy and set course directly for southern England. With inverted gull wings and, a huge fixed undercarriage, and dark paint these aircraft had a menacing, vulture-like look, especially from the front, which is how their targets would normally see them. Their dive bombing was highly accurate. Markings on the side windows of the cockpit indicated exact angles of dive, air brakes opened automatically to steady the bombing platform, and the elevator took over automatically to recover from a dive at a given altitude if the pilot should black out from pulling as much as 6G. The army would follow the Stukas and conclude the conquest. The Stukas were normally fitted with sirens that screamed as the aircraft dived, sowing panic among civilians and soldiers in their path; they had now been silenced because they caused drag and were less frightening to trained defenders.

The success and prestige of this elite Luftwaffe force had always depended on local air superiority. The Stukas did well against civilian populations or lightly defended troops but were far less effective against trained gunners on the ground, and when confronted by modern fighters they were sitting ducks, especially after bombing as their speed, already checked by air brakes, dropped off further in the steep recovery climb. Armed only with two forward and one rear machine gun they flew, at a mere two hundred miles per hour, into combat with aircraft which utterly outclassed them.

Squadron Leader Hope, having that day taken over as CO, led the waiting patrol of Hurricanes from Tangmere to Bembridge, then was ordered to orbit base at twenty thousand feet. 'You are only to engage the Little Boys [fighters],' ordered the controller; 'On no account must you attack the Big Boys [bombers].' This instruction violated Dowding's direction to avoid fighter-to-fighter battles and remains inexplicable to this day.

The Stukas came into sight, scores of spots in diamond formations against the haze. There were no enemy fighters, and the sky was full of targets. Hope reported the Stukas crossing the coast two miles east of Selsey – could he go for them? The controller repeated that he must only attack the Little Boys.

Hope was puzzled and frustrated. But he had no doubt about what he must do. The Hurricanes were at the north end of the airfield at twenty thousand feet, turning to port and ideally positioned to attack the Stukas, below them and to the south, presenting irresistible targets. When the Stukas put out their air brakes and began their dives, Hope defied the controller and 'after a good search I decided that we must attack the bombers'. As he dipped his wing he had a fleeting glimpse of a bomb exploding within fifty yards of his parked car near dispersal and felt all the more justified. Almost every Hurricane scored a kill or a 'damaged', and the surviving Stukas were chased out to sea. One showed remarkable aggression by attaching itself to the tail of a Hurricane and trying to shoot it down with its two machine guns; by contrast, another when menaced by Davis landed under control in a field between Pagham and Bognor and its crew surrendered to the Home Guard.

Sergeant Gray's combat report describes how he dealt with a Stuka when he had run out of ammunition:

> My first attack was from line astern quarter, but I was out of ammunition. The E/A dived closer to the water. I didn't receive any return fire from the rear gunner, so I dived again on the E/A from astern. I passed very close to it, and again it seemed to get lower on the water. I dived a third time and passed within fifty feet over the E/A. This time it was within fifteen feet of the water. When I turned again there was a large circle of disturbed water and no Ju 87 to be seen. I searched the area but could find nothing.

As Hope landed, and just as Jack Riddle was about to take off, a Hurricane crossed the airfield boundary trailing smoke or glycol coolant. It was Billy Fiske, whose aircraft had been hit, almost a fluke, by the lucky aim of a Stuka's rear gunner. He got his Hurricane down but it hit the ground heavily and burst into flames. He was pulled out of his blazing cockpit after some minutes with terrible burns and his clothes on fire by a nursing orderly, Corporal G. W. Jones, and Aircraftman Second Class C. G. Faulkner, who raced across the field in spite of the falling bombs. Both airmen were awarded the Military Medal. Within days Fiske's Hurricane was repaired and back in action. Kershaw writes that Fiske's Hurricane caught fire in the air and that although urged to bale out he said, 'I'm going to try to bring it back,' and that he attempted to do so with his clothes and legs on fire. In fact, what if anything Fiske transmitted is unclear; neither Hope nor Jack Riddle, witnessing his fiery crash-landing and listening to R/T transmissions reported hearing any such exchange; neither did they even realise the crippled Hurricane was Fiske's and it would be unheard of for a pilot to remain in a blazing cockpit to save his aircraft.

There was no denying Fiske's bravery. On the following American Independence Day, 4 July 1940, William Meade Lindsley 'Billy' Fiske III was buried at Boxgrove Priory, near Chichester (where it would remain neglected for nearly sixty years). The dedication was simple: 'AN AMERICAN CITIZEN WHO DIED THAT ENGLAND MIGHT LIVE.' The words were Churchill's and the ceremony, unusual for an undecorated RAF pilot, had the American public in mind. In this Churchill showed an astonishing ignorance of American public opinion, just as he later would of British. Fiske was held up as an example of what right-minded Americans should do: 'Here was a young man for whom life held much,' said Sir Archibald Sinclair, Secretary of State for Air. 'Under no kind of compulsion he came to fight for Britain. He came, and he fought, and he died.' It is unlikely, however, that sending their young men to meet Fiske's harrowing end would appeal to a largely uninformed American public, much of which saw no difference among the squabbling European powers, didn't think Britain could fight, mistrusted her imperialist motives, and would in some cases say, 'Let God save the King.' It was the heroism of *British* fighter pilots and the stoicism of the blitz victims that would begin to win American moral support.

Tom Waterlow, the adjutant, wrote a letter of condolence to Rosie, which ended with the tribute, 'I shall always look back at the time that Billy was with us as one when the squadron was at its peak and covered itself in glory. Billy played a great part well in that time and always kept everyone's sprits at the best. You will let me know both now and later if there is anything I can do.'

Flight Sergeant Harnden wrote:

> A very memorable day, as on this day Tangmere [was] visited by many Stukas or Ju 87 aircraft, the whole affair was amazing from start to finish. At 12.45 hours the squadron aircraft made a panic take off and went to 10,000 feet or so as instructed and then at 13.00 hours a large formation was seen approaching the drome from the south, no alarm

> had been received, then the Tannoy went, 'Take Cover, take cover', that was all, but in such a tone that everyone moved alright; anyway the Ju 87s made two attacks and carried out some very accurate bombing, they demolished three of the four hangars, 601 hangar being untouched!, the stores, sick bay, workshops and one amongst some Blenheims on the aerodrome. By this time the aircraft of ours up above had seen what was going on and came down after the raiders and caught the whole lot as they were going out to sea.

The sole medical officer at Tangmere, Courtney Willey, was buried under debris and deafened when the sick bay was hit, but assembled a makeshift facility by communicating with the staff by hand signals. He gave Fiske morphine, but afterwards worried that due to his own shock he hadn't marked Fiske's forehead indicating the fact. If the hospital to which Fiske was then sent had also administered morphine it could well have been fatal. However, Willey said he believed that Fiske was too far gone to survive. He was awarded the MC.

A German Intelligence Staff map summarised the results of the three-day assault. Using red for 14 August, blue for the 15th and green for the 16th, generous curves that conveyed nothing of the ferocious resistance encountered swept across the southern counties of England as far inland as the Severn, liberally punctuated with coloured triangles where airfields had been blitzed. A red triangle was awarded to Manston; a blue to Andover, Worthy Down, Kenley, Biggin Hill, Redhill, Rochester, Eastchurch, Lympne, Hawkinge and Tangmere. Tangmere's blue was outlined in green, signifying the two days' raiding.

Impressive as this looked, it totally misrepresented the position. Incredibly, some of the airfields selected for attention were not even used by fighters, while among those that were, Tangmere, despite its double treatment, didn't cease to be operational. A useful addition to the map might have been coloured blobs where aircraft had been shot down. There would have been 140 for German, sixty-three for British.

Despite the loss of life the damage to Tangmere was slight and didn't interrupt flying. Harnden wrote, 'As the attack was concentrated upon the hangars and squadrons were operating at dispersal, the operating efficiency wasn't affected in any way, in fact [it] was a very good demonstration of how well organised things were there'. Some installations were damaged, Hope's car was hit with shrapnel, and the old man who had sold newspapers on the camp was frightened out of his wits and never seen again.

Two airmen sheltering in a conduit built for the camp's central heating were killed outright by a direct hit on the boiler. Their bodies were being placed on a tarmac path as some captured German aircrew were marched through the gate. One of the Germans was imprudent enough to laugh when he saw the bodies, and a senior RAF officer who was walking to meet them lengthened his stride and punched the German in the nose. That evening the prisoners were given brooms and made to sweep the bomb debris in the hangars under the interested scrutiny of the station's airmen.

In a later engagement that day, Hope celebrated his appointment to CO by shooting down a solitary, already-damaged Me 110 that was careless enough to stray over Tangmere on its way back to France. The Messerschmitt fell in full sight of the camp's inhabitants into the grounds of a military headquarters one mile from the runway, blowing a crater that soon afterwards was turned into a makeshift swimming pool.

While collating the data brought back by its crews the Luftwaffe relaxed pressure on 17 August, but resumed the attacks with vigour on the 18th. The heavy loss of Stukas, sometimes as many as half an entire formation, was very bad news for Goering and very good for Dowding.

The squadron was due for a rest. The chain of reactions from 'scramble' to 'pancake' were the fabric of the pilots' lives; even waiting for the ringing of a telephone or the clanging of a bell was taking its toll on their ability to remain constantly alert and with quick reactions. During air defence exercises held before the war a reporter had written in an aeronautical journal, 'All the pilots were in agreement that the physical and nervous demands of the repeated flights under combat conditions could not be maintained for longer than a week.' This assessment had only been confounded by an immense effort of will and loss of efficiency.

Aitken's brief tenure as CO, followed by similarly brief ones of a month each by Squadron Leaders Hobson and Eddie Ward, then by Hope for only the last six months of 1940, plus the attrition of experienced pilots and their replacement by green ones straight from training schools, began to harm the squadron's cohesion and morale, and the ground crews noted this with dismay. An exchange with pilots from other groups, especially No. 10, which was chomping at the bit, was long overdue.

Although the island's defenders were nearing exhaustion, the British public was little disturbed by the cataclysmic events above their heads. On this fierce day of fighting, Kenneth Lee from another squadron baled out onto a golf course. 'There I stood at the bar,' he recorded, 'wearing a Mae West, no jacket, and beginning to leak blood from my torn boot. None of the golfers took any notice of me – after all, I wasn't a member!' On 19 August, in exchange for No. 17 Squadron, 601 was withdrawn from the southern battle skies to Debden, just north of the inner ring of fighter stations around London. Flight Sergeant Harnden's pride in his ground crew was bursting while looking after the Hurricanes of No. 17 Squadron: 'When they came to us, the pilots didn't seem to have any idea about getting away in less than two minutes, and were absolutely amazed at the efficiency and enthusiasm of my lads, which was certainly a compliment from a regular squadron, and I'm glad to say they were every bit as keen as us after a couple of days.'

Here the fighter pilot's life, though active, was more peaceful than over Hell's corner. Or so it had been. The very next day, as if by evil design, the Luftwaffe switched tactics and directed its attacks inland, to London and its environs. The overdue rest for 601 would have to wait.

For the next few weeks the squadron fought the enemy, destroying some aircraft, losing some of its own. It was during this time that the Battle of Britain would be won or lost. If the attrition rate that the RAF suffered in the fortnight from 24 August to 7 September were to be maintained another three weeks, its total fighter reserves would have been swallowed up. As the Luftwaffe increased the weight of its fighter escorts the strain on Fighter Command intensified. On 601 the aggressive instinct became stronger and began to eclipse that of survival. Flying Officer Tom Grier, after destroying an Me 109 in a 'twisted, smoking mass', found himself behind one Me 110 and in front of another, but 'stayed there presuming that the other one couldn't stay on my tail any better than the one in front could get away from me'. Pilot Officer Humphrey Gilbert was unable to conceal a certain disdain for the pilot of an Me 110, who when attacked, 'in my opinion deliberately and I consider quite unnecessarily force landed about one mile north of Shoreham'.

A report by Michael Robinson, who had been posted to Debden for 'rest' with 601, blended exuberance with the tenacity of a killer:

> Having attacked two Me 109s at the mouth of the Thames estuary, I saw a third 109 flying past me. I followed him down to ground level and chased him southwards. He never rose above 100 feet until well south of Maidstone and then throttled back. I overtook him and formated on him, pointing downwards for him to land. He turned away so I carried out a dummy quarter attack, breaking very close to him. After this he landed his Me in a field at about 140 mph some 25 miles south-east of Maidstone. The pilot got out apparently unhurt and held up his hands

> above his head. I circled round and waved to him, and as he waved back I returned and threw a packet of twenty Players at him and returned to base. He picked up his cigarettes.

There was a paradox in the daily lives of the pilots. By day they lounged in the crew room wearing the bulky yellow Mae Wests that were a constant reminder of hazards they might face at any time, playing shove penny or darts or reading magazines, or sat outdoors around a trestle table with tea brewed in a proper teapot and drinking from cups and saucers. Some of the others might be in the air. The ringing of the telephone would cause their stomachs to knot and they would stop everything to look for clues on the face of the officer answering. Had the NAAFI van broken down? Was there a change in runway direction? Was it an order to scramble? If scrambled they would race to their Hurricanes, where ground crews would be waiting to help with the straps and pull the chocks away. The tension would diminish as they raced to their aircraft. At night, unless they were required to camp at readiness in the crew room, they might be in proper beds, those with wives perhaps in rented cottages or houses. They were determined, but not bloodthirsty, and everyday courtesies were not overlooked. They were defenders of their homeland, and their duty was to destroy the enemy at whatever risk to their own lives. It wasn't difficult for No. 11 Group, especially 601, to understand what they were fighting for. They flew over it all the time. This was their territory, with Tangmere and The Ship and Porte Lympne and Botolph's. And they knew what they were fighting: Hitler's speeches, grotesque racial theories and barbarous acts were no secret. Now, having invaded its neighbours, Germany wanted to do the same to Britain.

Death had become a familiar visitor. It would be mistaken to assume that the pilots, for all their nonchalance, were blind to the dangers, equally so that their wives underestimated them or lived in constant and debilitating anxiety. Michael Duke Doulton of the famous English ceramics company, at six feet eight reputedly the tallest pilot in the RAF, was shot down over the Thames Estuary in a fierce dogfight on 31 August. He had joined No. 604 (Auxiliary) Squadron in 1935 and met Carol Christie, an American, while skiing in Davos in 1938 and married her in March 1939 at St Mary's, Shrewsbury. In June 1940, just in time for the Battle of Britain, he transferred to 601. Two days after Michael was shot down Carol wrote a poignant letter to her father in America. Calmly she poured her heart out. Because it opens a rare window into the mind of a fighter pilot's wife and the tightly-knit, intertwined lives of pilots and wives of 601 it is worth quoting in full:

Dearest Father,

By this time you may have a cable. If you have not any cable from me, it will mean that Michael is safe. But, although I shall wait another forty-eight hours before telling his family and leaving here, I am quite sure he is dead. I shall cable you tomorrow or the day after that he is missing and that I am coming home to have my baby.

You have never been able to believe that I have been happy in the middle of the blitzkrieg, so I doubt that you can realise now that I am quite calm and full of hope. Not for Michael. Yesterday afternoon Molly Riddle had just arrived from Essex, and the squadron was due at any moment. Michael was to have gone on leave at one o'clock Saturday Aug. 30, but waited over to come down with the squadron. It was typical that he should be killed while working overtime. As Molly and I were talking, in came Tom Waterlow, the adjutant. We knew at once that something was wrong. He said that he had bad news for me – Michael was missing. Saturday afternoon there was a tremendous battle over the Thames Estuary (near Birchington, I suppose, where we were so happy) and four of Michael's flight were shot down. The other three came down by parachute on land on either side of the Estuary (40 miles across there) but Michael hasn't been seen or heard of. By now, if he had come down on land, we would have heard, unless he is unidentified; but after 48 hours, I think it is unlikely. I pray that he came down into the sea and that he will just vanish. I have no desire for the horror of bodies and funerals to come between me and my last happy memories of Michael young and strong and confident. To have him just disappear suddenly and cleanly in the midst of life and never return broken and dead is how it should be. Perhaps I ask too much. You see, ever since the war started and I expected to lose Michael, I have wanted a child. So that if Michael were lost, something of or love would live on, and so it wouldn't be the end of everything for me. When Copper Prescott was killed in November, I remember Imogen's calmness. Her baby was born in February. Now what is behind me is all happiness, and no two people were ever so happy right up to the last minute as we were, and I don't look forward to complete emptiness. I have his child to think of, and even if I felt hysterical I wouldn't be, because of the baby. I wouldn't leave England otherwise, and baby must have the advantage of my American birth, and be born without fear of bombs. The only thing is, of course, that I can bring no money. But I still have some AT&T and shall try and bring all the clothes for it that I can. Unfortunately, as

Michael is missing, I probably shan't be able to get any life insurance, etc. before I go. But it will wait for me here when I return after the war. Michael's child must be an English child.

Tom was marvellous and didn't try to raise any false hopes. Every hour that goes by makes it more certain that he is gone. About an hour after Tom was gone, he came back again, bearing a bottle of champagne sent by Michael's flight commander [Willie Rhodes-Moorehouse]. He is a crazy young man who loves the pleasures of this earth and isn't ashamed of it, and is a glorious pilot. He admired Michael immensely and got a great kick out of him. Only Willie would have thought of sending me champagne. Mrs Chapman from the house (a grandmother since Thursday) and Molly and I drank it, and drank to Michael and the baby. Michael would have been so amused by it and so happy. It was a good thing and I slept soundly all night, and for once there were no bombs and no gunfire. Today we have been on a picnic on the Downs, and Jack [Riddle] is coming tonight. I have told Eleanor and arranged to go here as soon as Tom tells me all hope is gone. I have none; and I am happier that way. But I shan't tell Mimi [Michael's mother] until a few days pass. She is staying with Sherwood and he must tell her. I am not going to see any Doultons for a bit. I will have no wake.

How many times Michael and I have said: 'If we died tomorrow, how much more happiness we have had than most people.' I am now finding that I meant it. I keep looking at his things – everything reminds me of some happy time together. The little bear he pulled out of his cap at the station when he left Davos. The bluebird pin he gave me to keep me happy while I was at home. The five tins of his favorite oxtail soup sitting in the cupboard. His brushes. I don't want to put them out of sight – I want to have them around and relive every glorious moment. And the thing I dread most is telling his family. I am calm and happy now with his things and Molly, but gloom and misery will merely infuriate me. I can't dictate to other people how they must feel, but to me it is wrong and wicked to mope and to say all the silly things about why, and 'he was so young' etc. Anyone who knows Hitler knows why, and what better way to die. It was Michael's way of dying. And it is a fine way. I am selfishly going to avoid all ghouls and weeping relatives for a while. I can give them more details than I have just given you. Want to be with Eleanor or his young friends who understand him and will talk about him and remember him alive, not brood over his death. He was, with one possible exception, the finest man I ever knew, and just because he isn't here to put his foot down I am not going to let people talk of him like a ghost. And I won't be treated like Banquo's ghost either, and have people stop talking happily when I come into a room. I have told the Chapmans at the House [owners of the house rented by the Doultons], who have been so kind to me, and Mrs Chapman wept her heart out. She came over all tears and kisses and dreariness and we gave her a glass of champagne and got the conversation around to Michael, and before she left she was laughing and talking about him naturally. The day before he died he wrote one of his typical sweet notes to her, thanking her for her kindness to me.

Michael was so happy about the baby – thank God he knew before he died. He made me promise I would go home to you if he were killed. But even though we naturally had to face the fact that he might be dead the next day, we were so used to the idea after a year of it that it didn't depress us. That's what may be difficult for you to understand. I am doing exactly as I had planned for so long. From Mother's death, I remembered how hard it is to make decisions at a time like this, so I had a schedule planned of what I would do. And I am falling into it easily. And yet the knowledge that [I] might lose him, after the first weeks of the war, when the idea took getting used to (Mimsie [Carol's younger sister in America] will tell you that) hasn't made us any the less happy. We weren't feverish about it. We just loved each other so that we had no time to worry. I can't believe that two people were ever so close. But you know that.

When I come home you will be unhappy. You loved Michael (Michael was so sorry he never had had the time to answer that beautiful letter you wrote him when we were in Kent) and you will grieve for him, and it will hurt you to see me come home like this. But don't waste grief on me. Remember that I have a child to look forward to and that I have had eighteen months of wonderful happiness. And I am too proud of Michael to be unhappy in a petty way. And I am young and healthy – I have not felt better in my life before; and the baby agrees with me.

I shall write Mimsie, but send her this. She will feel it very much; she adored Michael as he loved her too. I shall miss Mimsie so far away this winter, but don't let her feel she must stay behind. She must go on with her life and when she comes home for Easter vacation, she will be aunt to Paul Michael or a Miriam Hope [if a girl]. I am a bit hazy about when the baby is due, but April Fool's Day is a safe guess. You can tell people about the baby. I don't want them to think I am running from

England and danger. And tell everyone most emphatically not to avoid talking about it, or to fall on my neck and be sentimental. Tell them I'll probably bore them stiff singing Michael's praises just as I did when we were engaged.

I feel that I am so lucky, not only to lose him in such a glorious way, and to have this child to look forward to, and to be able to have the child in safety, but also to have the finest, most understanding father that ever lived. Feel that I understand you better than when I was living with you, as you are like Michael in many ways, and through him I have come to appreciate you. And I shall always be grateful that you didn't try to make me come home before I was ready.

Much love,
Carol

P.S. When school opens, will you have Mr Davidson say a prayer for Michael in chapel.

The ground crews felt it too. They were part of a team, bound by warm ties of mutual respect to 'their' pilot. Flight Sergeant Harnden had the opportunity and emotional distance to write:

This was a great loss for the squadron as Doulton was absolutely brim full of enthusiasm and efficiency, he was in 604 in the peace days and the early days of the war, but came to us while the Blitz was on because of the inactivity in his own squadron, well over 6ft tall. One could always tell when the sqdn had been suddenly called to readiness by him streaking round the tarmac on his bike well in advance of the tender. He was always exceptionally polite, I can remember him now, 'Well Flight, do you think we could –', never once did he give anyone a direct order to my memory, and a lot of the gen in these notes are due to the remarks he made in the pilots authorisation of flight book. Yes, he was a real loss to the squadron.

It was dawning on the Germans that they too were suffering horrifying losses in aircraft and crews. Whereas a British fighter destroyed might mean the loss of a valuable pilot who might crash-land or bale out and be able to fly again, a German bomber shot down over British territory meant the loss for good of a crew of five. This was Dowding's strategy. He was destroying more aircraft than he was losing, and many times more aircrew since the bombers had several crew members. Even so, RAF pilot replacements were not enough to compensate for losses, and continued attrition would favour the Luftwaffe. The Germans did not know that three more weeks at the same rate to both sides would spell doom for the RAF.

Dowding had decided after 18 August to pull squadrons that had suffered 50 per cent or more losses from the front line for rest and refit, and 601 was moved from the London area altogether. On 6 September, the day before the transfer to Exeter, there was a final 'party' with fifty German fighters over Essex. It was one of the fiercest ever. Two 601 pilots were wounded and baled out and three others force-landed. Two were killed. One was Raymond Davis, the squadron's other American, the second one Willie Rhodes-Moorehouse. Both pre-war Legionnaires and recipients of the DFC, since the Borkum raid they had thrown themselves into every stage of the squadron's war effort with the same verve as they had into the peacetime mischief. Davis was shot down by an Me 109 over Tunbridge Wells, crashing inverted into a garden. The civilian who first reached his body wrote to his wife, Ann, Hope's sister, to apologise for intruding but said he had been in the First World War and wanted to reassure her of one thing: he could tell that Davis had a received a bullet [or possibly shrapnel] to his head and must have died instantly. Flight Lieutenant Carl Raymond Davis, DFC, had nine victories to his score.

Flight Lieutenant Willie Rhodes-Moorehouse, then aged twenty-six and also with a record of nine kills, had only six days earlier been awarded the DFC. Accompanied by Amalia, who was still grieving from the loss of her brother Dick Demetriadi, he had attended the ceremony at Buckingham Palace where the King pinned the medal on his chest and spoke a few kind words while a military band played softly in the background. Willie had been a mainstay of the squadron; the one who bought the petrol station, who always paid the bills, who was in the Hurricane flight escorting Churchill to Paris, who was always there. Amalia declined several offers of marriage and died in 2003.

There is no record of how the pilots took these catastrophic losses among comrades and in many cases relatives. They didn't write or speak of fear but kept it to themselves, and if they had wives they would speak lightly of the risks. Apart from laconic entries in log books they left no written accounts. Interviewed after the war the survivors were blithely matter-of-fact. But the strain was building. Tension was becoming normal and feelings were hardening. Geoffrey Wellum, an RAF pilot in Bushell's No. 92 Squadron, must have been speaking for all of them when he wrote in *First Light* of sitting in his Spitfire cockpit waiting to be scrambled and thinking, 'I wonder which of us will be killed today.' As Tom Hubbard put it: 'We all knew it was like a game of roulette, backing black all the time. Our luck wouldn't come up forever.' Hubbard survived the Battle of Britain; what happened to him thereafter isn't known.

Flight Sergeant Harnden described the shock and dismay felt by the ground crew at the death of Rhodes-Moorehouse:

> F/LT Rhodes-Moorehouse (Willie to us all) failed to return, it is impossible for me to try to explain the effect this had on the Flight, it seems so absolutely impossible that this smiling, undaunted person was gone forever, I just couldn't believe it, and all day we just prayed for news of him. It was some time before the wreckage of 'G' was located but it was finally found on a sewage farm at Highbroom, Tunbridge Wells, in Kent, an absolute wreck as it apparently dived straight in, so how he die[d] we shall never know, but his example will always live with 601. The times I remember most vividly were waking him for early morning readiness at somewhere about 3 o'clock, always 'Allright Flight, call me again in five minutes'; and then taking off, how he would be doing up the strap of his helmet whilst opening the throttle & taking off; he was a born leader and a marvellous personality. Mrs Rhodes-Moorehouse, his mother and his aunt, Waterlow introduced me and I found all absolutely charming, my those people can take it on the chin, just think his wife had lost her husband and her only brother, Dick Demetriadi.

Next day, 7 September, when a much depleted No. 601 Squadron was put on the 'C' List and posted to the former aerodrome of Whitney Straight's at Exeter, Tom Waterlow confessed to having no regrets at seeing the squadron pulled from the front line, if only temporarily. 'I don't enjoy my work', he said, 'when it entails rushing around the countryside looking at crashed aircraft and identifying my friends by the numbers on their machine guns.'

That same day, partly because they were unable to solve the problem of daylight fighter escort, but mainly in retaliation for a raid on Berlin that was in turn a retaliation for the accidental bombing of part of a London, the Luftwaffe changed tactics to the night bombing of London. This tit-for-tat was due to a Luftwaffe crew's mistake. A Heinkel bombing the legitimate target of RAF Croydon had also accidentally hit the suburban area, which was part of greater London, in contravention of Hitler's orders. Churchill, feeling compelled to respond, ordered an attention getting if ineffectual raid on Berlin. This astonished and enraged Hitler and Goering, who felt in turn compelled to mount the large-scale bombing of British cities. The long feared nightmare had come about, and it was Britain's salvation.

The realities of total war were now brought home painfully to the people of Britain. More than one million London houses were destroyed or damaged and more than 40,000 civilians were killed, almost half of them in London. But the Blitz was a godsend for Fighter Command. Now its aircraft and pilots would be saved from total attrition and, with the advance towards October with its Channel weather making an invasion no longer feasible, this meant the true Battle of Britain ended with the beginning of the night bombing. Events proved the dire prediction of Stanley Baldwin that 'the bomber will always get through' to be correct if he meant some bombers and not all. Total defence against bombers by day and night was clearly not possible, but the conclusion he drew from that, that there is no point in having fighters, only bombers, was proved dead wrong.

The new Luftwaffe policy, though hard on the citizens of London, was a blessed relief. Although no one yet knew it, the Battle of Britain had been won, and Britain had won it. The material factors had their genesis in the Air Ministry decisions and great technological advances of the 1930s; radar, and the eight-gun, single-seat fighter; the human factors were the blood, toil, tears and sweat Churchill had promised. The nation paid for its victory in its best blood, and an official RAF historian wrote, 'When the details of the fighting grow dim, and the names of its heroes are forgotten, men will still remember that in the summer of 1940 civilisation was saved by a thousand British boys' (Denis Richards, *Royal Air Force, Volume I – The Fight at Odds*).

Flying Officer Barnes, 601's intelligence officer (IO) at Tangmere whose job it was to interrogate the pilots after they returned, smoke streaks on their wings indicating they had fired their guns, tugging off oxygen masks before answering his standard questions, often craving a cigarette or a pipe, also noted those who didn't return. The respect and affection he shared with the ground crews for the pilots is evidenced by a poem he composed and left behind upon being posted away, titled 'Goodbye!':

Good luck, good hunting, six-o-one,
From one whose task is nearly done.
No more my daily round begins

With flapping bumpf and drawing pins;
No more I stand and theorise
Whilst you are at it in the skies!

No more I ask you, long and loud,
'What is the tenthage of the cloud?'
No more will waving hands explain
The antics of the aeroplane,
And warm my heart as they declare,
'We missed him here – but got him there!'

No more will my reluctant pen
Shatter gallant records when
It seems as if it much enjoyed
The word 'damaged', not 'destroyed'
Good luck, good hunting, six-o-one!
You're shot of me – my time is done,

But I shall miss the quick surprise,
The sudden thrill of smoke-trailed skies,
The quick relief when decent types
Come duly back – and light their pipes,
And try, with diving rolling hand
To make their IO understand!

When days are dull, or not so dull,
Activity or boring lull,
'Huns in flames!' or 'standing by'
For 'Liz' the loping butterfly.
Fog on the deck, or clear blue skies,
I'm going to miss you – blast your eyes!

And if you see, your sortie done,
Two one-one-Os instead of one,
You can believe in what you see;
The second is the ghost of ME!
And when you have a lucky day,
And one-o-nines become your prey

You'd be surprised if you could hear
How very loud a ghost can cheer –
'Good-bye then now, and thank you boys!'

The most intense and violent period of air warfare of all time ended. It is sobering to realise that if the Germans could have used droppable long-range fuel tanks to increase their Me 109s' combat endurance over England beyond the typical ten to twenty minutes, by sheer numbers they would have overwhelmed and defeated the RAF. But they didn't, and it was Hitler's first reversal, a balm to the British public's sense of shame for the ignominious evacuation of its army at Dunkirk, and perhaps above all a message to the world and especially the United States that Britain would never yield.

The Germans knew an invasion army would require barges and ships, and the navy knew it would need air superiority to protect it from the Royal Navy. Here the much vaunted Luftwaffe suffered an unexpected and shocking defeat thanks to the pilots of Fighter Command. Serious historians have been unequivocal in giving credit to the strategy of Dowding and Park.

The reality of air warfare had seemed other-worldly, even sporting, to the people of Britain, watching from the ground in the fine, un-English summer of 1940, barely disturbed from their shopping, gardening, or golf except by falling spent cannon shells, the occasional crashed aircraft, or a pilot parachuting down, British or German. They listened avidly to daily scores, usually inaccurate, of RAF and enemy aircraft destroyed. In the south of England people watched with fascination the arcs and circles of condensation trails etched against a blue summer sky, in what appeared to be a free show of British skill and courage. So remote was the action, so dashing and heroic the image of the fighter boys, it may have been easy to exclude thoughts, or more likely difficult to imagine, scenes of horror, pain, and death five to six miles above the earth. The threat of invasion, of which people had been reminded by anti-glider obstacles in open fields, the removal of road and city signs, warning signs everywhere, and the evacuation of children from major cities to the country, dissipated with the daily trumpeting of RAF victories.

No. 601 Squadron entered the Battle of Britain with twenty pilots, but within mere weeks had been reduced to nine with eleven either killed or wounded, including most of those representing the cream of British, Australian, and American society. Most of the old guard who had been drinking and tonics just a few months before were gone, either killed, wounded, or promoted off to lead other squadrons, where most were later killed, missing, wounded, or captured. The replacements, their flying courses barely completed, came and went so quickly, often after just a few sorties, that there was no time to get to know names and faces. Out went birth, wealth, public school and university, and accomplishment in sports as qualifiers: pilots were badly needed. Though not fully recognised at the time, an unstoppable social change had begun that would gather force throughout the war.

While the RAF was fighting the Battle of Britain, another battle raged within it. Dowding and New Zealander Keith Park, AOC of No. 11 Group, had pursued the strategy of avoiding enemy fighters and concentrating on the bombers, and doing so as much as possible over British soil where the RAF's pilots could be rescued and returned to the cockpit while at the same time challenging the enemy's flight endurance. No. 12 Group further inland, commanded by AOC Leigh-Mallory and egged on by the famous but egotistical squadron leader Douglas Bader at Duxford, who had lost his legs in an aerobatic accident, was urging 'Big Wing' or mass defence formations of several squadrons. Such massive formations would take time to assemble, risk higher losses, and reveal Dowding's hand – that he had reserves.

Dowding's ruse had been fully justified on 15 August with the resounding defeat of mass enemy attacks. The Germans assumed all reserves were being used and seeing only moderate numbers of British fighters grossly underestimated Fighter Command's strength. Dowding's artfulness in reserving Nos 12 and 13 Groups until they could be unleashed on unprepared attackers at the crux of the battle had turned the tide, both materially and in its psychological effect on the mystified enemy. While history declares Dowding and Park the winners, politics did not. Dowding's rivals closed in on him, including Sir Sholto Douglas, Deputy CAS and a Big Wing proponent (and after the war Honorary Air Commodore of No. 601 Squadron). Dowding was eased into retirement and replaced by Sholto Douglas, while AVM Park was sidelined to the Middle East, where he would later again show his supremacy as a fighter commander in a challenge equal to the Battle of Britain. The injustice to Park was real. However, the notion that Dowding's forced retirement was a shameful act proving that no good deed goes unpunished is less than fair. He was old and had long been scheduled to retire but had been retained so that he could finish the job. Although his retirement wasn't handled gracefully by an out-of-touch Air Ministry, his performance during the Battle of Britain was later described by Churchill as, 'an example of genius in the art of generalship'. Still, his time was up.

The rich and privileged part-time flyers of 601 had shown a high concentration of physical and mental courage, fighting spirit, and skill, and were highly regarded by the 'erks'. Questions do arise, however, about Fighter Command's priorities. More attention had been paid to the mass production of superb aircraft than to the needs of those who flew them. Why were Channel rescue facilities so inferior to those of the Germans, in fact almost non-existent? Why was there not there even something as simple as dye markers to drop on downed airmen bobbing in the water and on the brink of hyperthermia; had nobody read Hope's combat report after Mayers parachuted into the Channel: 'It would have been helpful to have some kind of marker …'? Why, even, did pilots not have practical flying overalls with pockets for maps and survival equipment? Saving expense on pilots for the sake of more machines was a false economy. Aircraft could be turned out much faster than pilots, while a pilot lost meant an aircraft lost but the reverse didn't necessarily apply. The senior RAF officers responsible for this imbalance also insisted on the shockingly hazardous Vic formations and formula attacks which cost many lives.

On June 22 1941 the Germans launched Operation *Barbarossa*, an invasion of Russia so unexpected that Joseph Stalin, who never trusted anyone, believed the assurances of Hitler, the world's most notorious liar, and forbade his commanders from taking defensive measures. Some revisionists have argued reasonably that Hitler's heart was never in the invasion of Britain, that he was impatient to attack Russia before the United States entered the war, that he considered Britain beaten and would soon be ready to do a deal. They add that regardless of Hitler's true intentions the obstacles to an invasion, mainly the Royal Navy, were too great. Perhaps. Although it may equally well be argued that in the absence of fighter cover Britain's warships would have been sunk by the highly accurate Stuka dive bombers, we will never know, and it is irrelevant. The War Cabinet, Fighter Command and No. 601 Squadron were not going to allow the Stukas to bomb their convoys and airfields unmolested and their Messerschmitts to roam destructively over the homeland. The fact is that Hitler tried and Hitler failed.

Chapter 7

Whitney Straight's War

Closed. All enquiries should be addressed to the United States Embassy, Berlin

Notice on door of the United States Embassy in Paris

Flying Officer Willard Whitney Straight was twenty-seven years old when he was called away from 601 Squadron in 1940 for a secret mission, at his own request. He had joined the Legion shortly before the war for the simplest of reasons: not to fly or have fun or extend his social life but because he loathed the Nazis and saw this as a way to fight them. He took the war seriously, personally. Flying was a natural path to his war against Germany and 601 was natural too, but it was a means, not an end. At meetings with Nazis in Berlin Straight found their diplomats bizarre and haughty in the mould of the odious foreign minister Von Ribbentrop, who wasn't even a real 'Von'. Their architecture was sinister and intimidating; the skiers he met struck him as arrogant. It was all very ominous.

There was some cause for the Nazis' bluster. They had a highly efficient military including an air force that was modern and respected. Churchill had said that 'Germans are always either at your feet or at your throat', and so it seemed. They hadn't lost the Great War, they had been betrayed. Invading neighbouring countries was what Straight would have expected them to do, and Britain was sure to be in their sights. Perhaps more serious than his squadron colleagues, Straight saw flying primarily as a means, but not his only means, of waging war, and he was impatient to begin.

He had already lived a full life. By age sixteen, although too young for an 'A' licence, he had achieved over sixty hours of solo flight. He became a famous Grand Prix motor racing driver while an undergraduate at Cambridge, formed his own motor racing team and led it to victory in the South African Grand Prix; He designed and operated airfields throughout Britain and helped design Miles Whitney Straight aircraft.

With deep-set eyes under bushy eyebrows on a well shaped head from which the hair was beginning to recede, he had a pleasant, deceptively gentle face, which belied great will, physical energy, and cool nerve. When war came he was one of the most junior pilots on the squadron but almost one of the most experienced, and this tended to set him apart from his comrades. The 'Phoney War' made him restless. The war demanded his full personal involvement.

On bumping into an Air Ministry friend in Piccadilly he asked whether 'a nasty job, somewhere' could be found for him. The plea brought a swift response and within a week, while Straight was sitting in his Blenheim on night readiness at Tangmere, a teleprinter message was handed up to him in the cockpit ordering him to report to the Air Ministry next morning.

At the Air Ministry conference, attended by senior RAF officers, Straight learned the nature of his assignment, and it didn't involve flying. The Germans were expected to invade Norway in anticipation of Allied interception of their essential imports of high grade iron-ore from Sweden, which supplied two-thirds of their needs. The plan was originally Churchill's as Lord of the Admiralty, but the government had dithered for months because the plan meant invading a neutral country. Now the Germans were up to it, and this called for quick action. The Royal Navy would mine the waters around Narvik while a commando force was sent immediately to locate airfields for incoming RAF fighters, which would have to take off from naval carriers. Straight would be part of the commando force. His job was to land with them by sea on the coast of Norway, locate frozen lakes that could serve as airfields and, as the commandos left, stay behind to blow up the place if the Luftwaffe arrived.

Straight went directly from this conference to his lawyers, made out a will, then returned to the squadron to await sailing time. When called away and without being able to explain why to Loel Guinness the CO he left Tangmere and boarded HMS *Devonshire*. But the mission was stopped before it could sail by a high priority signal, which was rushed to Vice Admiral John Cunningham, who in turn dashed ashore to revise plans with the Admiral Commanding Rosyth. German forces had already sailed for Norway. The British cruisers rapidly decanted their lodgers and joined the Home Fleet in pursuit of the German navy.

Back at Tangmere the pilots began to rib Straight about his cryptic comings and goings. 'Lining up a soft job somewhere, Whitney?' they asked provocatively; 'There'll be plenty to do here pretty soon.' Straight revealed nothing and continued with the convoy patrols until called once more to the Air Ministry.

The Germans were now in possession of all the ports and landing grounds in Norway, and something had to be done to provide bases for aircraft to support the projected landing there of British troops. The only possibility was frozen lakes. Straight, with his civilian experience of airfield construction, was to find suitable lake surfaces from which Gladiators and Blackburn Skuas, to be flown off carriers, could operate. It was a far-fetched plan, bred out of desperation. He assumed the temporary rank of squadron leader and reported to the commander of *Sickle* Force at Rosyth.

Conditions were close to chaotic on board HMS *Bittern* at Rosyth in Scotland on the evening of 14 April 1940, with marines loading guns and ammunition in a furious attempt to sail at midnight, but even at three in the morning the loading was still not finished. Straight found the atmosphere so exciting that despite the cold and damp he was unwilling to tear himself away from the scene for several hours, but when eventually tiredness overcame him he looked for a comfortable place to sleep and settled on the captain's bunk.

There was a heavy swell next morning, and as the ship rode at anchor many of the seamen were being sick. The troops were so hopelessly incapacitated that the departure was postponed by a day. The captain of Bittern repossessed his bunk and Straight found a camp bed for the night, transferring next morning to more comfortable quarters on HMS *Auckland* just before the fleet sailed for Norway.

On the morning of 17 April Straight awoke with a feeling of great excitement, and after bathing and shaving went on deck in his life jacket to find the gun crews fidgeting at their stations and the captain chain smoking cigarettes. Aandalsnes, where the party was to land, was less than a hundred miles away, and news had been passed by a Coastal Command aircraft that a U-boat lay in the mouth of the fiord which had to be entered. A destroyer steamed ahead at full speed to discover whether Aandalsnes was in enemy hands.

The destroyer reported that Aandalsnes hadn't been taken and the coast was clear. At six that evening the convoy approached the snow-blanketed peaks, three cruisers covering the entrance to the fiord as the four naval sloops sailed in. At eleven o'clock Straight went ashore, in the ghostly light of an arctic moon, accompanied by two naval officers, Captain Allen and Lieutenant Smith. The three held an immediate conference with the local British army commander. There had been a British landing at Narvik a few days before and the high command was now established, though tenuously, at Donbas. They decided their first duty was to join the High Command, and set off in two cars, each with a rifle, fifty rounds of ammunition, a revolver, and food for two days. They were repeatedly halted along the rough road by

Norwegian sentries for there was no certainty where German paratroops had landed.

At half past one in the morning they roused the Colonel from his bed at a regimental headquarters and demanded to know how they could reach the C-in-C at Donbas. The Colonel replied that he wouldn't expect them to break through the cordon of German troops ahead without a field gun. Straight learned that there was a field gun at Aandalsnes and with Allen and Smith he returned for it. After they had collected the gun, however, Straight received a telephone message to forget his attempts to contact the C-in-C and to begin immediately the search for frozen lakes. He took his leave of the naval officers and motored off on his search.

Lake Lesjaskog appeared promising. It had a large enough surface but was blanketed with three feet of snow. Straight found the local mayor and together they mustered a large labour force of two hundred men, all civilians, which had soon dug a road to the lake and began clearing the snow.

For the next few days Straight lived in circumstances so strange to him that he found it hard at times to credit that he wasn't dreaming. At frequent intervals during the long and eerie arctic day, with its luminous mountains and sombre sky, enemy aircraft hummed overhead with their cargoes of men and supplies from Germany to Trondheim a hundred miles north. On the ground British sailors erected Oerlikon guns in hasty preparation for the bombardment that was sure to come. Close by, Norwegians fought German paratroops in isolated skirmishes, the latter employing some distasteful methods to eliminate resistance. Straight wrote in his diary: 'Thursday, April 19th. Yesterday two Norwegian troops who were trying to help a wounded German, who was only pretending, were machine-gunned by another hiding behind a tree. I hate the Germans and their dirtiness.'

The Norwegian population was hardy, teetotal and almost embarrassingly keen to help Straight, who found himself the local hero and man of mystery. Not warlike by nature, these people were now in a mood to continue the fight, though their army had no central organisation and communications were not only poor but dangerous because of the spies and fifth columners tapping almost every wire.

Lake Lesjaskog was cleared, and Straight prayed that the faint sunlight wouldn't melt the top layer of ice. But he had still to find other landing grounds, and early in the morning he climbed on skis for over four hours, accompanied by four soldiers and a guide, to examine two further lakes east of Lesjaskog. His bitter disappointment at finding them deep under snow and inaccessible was only partly recompensed when, in the descent down a narrow, winding track, he discovered that he was the best skier, leaving his companions and the guide well behind.

Another possibility was Lake Vangsmjosa, to which Straight flew in a Tiger Moth fitted with skis, accompanied by Captain Øen, chief of the Norwegian air force. Having no helmet or warm flying clothes he was bitterly cold and in real danger of suffering from hypothermia. But they landed the little machine successfully on the lake at Vangsmjosa amid the cluttered remnants of the Norwegian air force, and Straight was able on his return to signal that this lake would be suitable for two squadrons.

Of the two lakes it was Lesjaskog that the RAF chose. On Wednesday 24 April, the first machines arrived – eighteen Gladiators of No. 263 Squadron, which had taken off from the carrier HMS *Glorious*, 180 miles off-shore, led by a navigational Blackburn Skua of the Fleet Air Arm. The Gladiator pilots, commanded by the valiant Squadron Leader Donaldson, had the challenge of making their first carrier take off in a snowstorm, and their first landing on ice, separated by a 180-mile flight over water with inadequate maps or navigational aids. They all made it.

So far the combined endeavours of Straight, the airfield labour force and the carrier-borne biplanes amounted to a spectacular success. Unfortunately, there was more involved when it came to establishing effective fighter operations than the selection of bases and the flying-in of aircraft. It was a gallant attempt, but it came to nothing, and the events of the next few hours were catastrophic.

'I expected trouble', wrote Straight in his notes, 'and it started early'. At six o'clock in the morning waves of unescorted Ju 88s and He 111s unloaded their bombs on Lesjaskog. Hearing the explosions, Straight rushed to the lake. The Oerlikon guns were firing hard, but not a single Gladiator was in the air; their wheels were locked to the ice, and frost had put the carburettors and control surfaces out of action. None of the pilots or mechanics remained by the Gladiators as two had already gone up in smoke, and within the next few minutes three more were destroyed on the ice.

As the days passed the German bombing intensified. The port of Aandalsnes was rendered virtually useless while Lesjaskog was reduced to splinters of ice and crippled or burned-out Gladiators. The Norwegian expedition was by now doomed, and on 30 April GHQ decided to evacuate the entire British force before it was too late. Despite the wisdom of such a course, Straight had deep misgivings about leaving the Norwegians since the evacuation had to be carried out secretly. He was saddened to see the

Norwegians still working hard to repair the runway as the British contingent quietly withdrew.

Straight reached Aandalsnes that evening. There was heavy bombing, which blew up the pier and left houses blazing all around what remained of the port. The sloop *Fleetwood* arrived to take on board the evacuees, but for over an hour an intense air raid prevented any movement, and Straight and his colleagues took shelter under the dock, sitting on a wooden beam with their feet inches from the water. When the bombing had subsided, Straight mustered the men who would have to wait until next day for evacuation and they retired for the night to the relative safety of a nearby village.

In the morning he returned to Aandalsnes with a corporal driver. Everything was unusually quiet, in welcome contrast to the previous night. For the first time in many days Straight began to feel a real elation, and in his relaxed mood he paid little heed to the sound of an aircraft that grew from nothing until it was almost overhead. He heard the bomb singing through the air and remembered saying to the corporal, 'This is going to be rather close!' There was a terrific shock; after that it felt like going under gas. There were dreams – pleasant ones – and the sapping of consciousness.

Slowly Straight became aware of the outside world and remembered what had happened. So he wasn't dead. How badly injured was he? His face was in mud and the pain shot down his back as he tried to move. He tried his muscles, deliberately and painfully, as though he had never possessed limbs before, and after ten minutes he seemed to be in command of his legs and managed to rise shakily to his feet, feeling warm blood running down his back and trickling over his greatcoat collar from the side of his head. Instinctively he looked for his cap but couldn't find it, then realised it hardly mattered. He made his way unsteadily to the lorry and found the corporal safe but shaken, and ordered him to drive on to Aandalsnes.

At Aandalsnes the bombing had been resumed as the Germans pursued the retreating force ruthlessly. Wing Commander Keens, in charge of the RAF elements, emerged from the ditch where he had been taking cover and walked towards their vehicle. To Straight's horror, he came to within arm's length and asked, 'Where is Squadron Leader Straight?' and his first reaction was, Christ, my face has been blown off. Then he remembered the mud. The bomb that threw him out of the lorry had been close. On the floor of a Norwegian dressing station he passed out again, and after his wounds had been dressed was transferred by whaler to a cabin in HMS *Sikh*.

The Norwegian venture was a disaster. The weather favoured the side with local air supremacy, and the Germans had it. The Gladiators, in any case hopelessly outdated, were virtually impossible to service under arctic conditions. Lesjaskog was eliminated as a fighter base, the machines abandoned, and the British pilots returned to England by sea. On the voyage back to England, with portholes closed and inadequate ventilation, sleep didn't come easily to Straight, and when it did was accompanied by nightmares in which he walked away from Aandalsnes looking for a boat, but could never find one, while the rocky hillsides, which made walking an agony, were according to his diary 'thick with lost and screaming men, fleeing they knew not where'. It was little wonder there were nightmares, for wounded army officers on the *Sikh* had some appalling recollections of the rout; of men who had lost their nerve, of officers in pain who had shot themselves, and Straight's judgment was recorded in his diary: 'It was maddening to hear the report of Chamberlain's platitudinous speech, i.e. enemy losses greater than ours, our losses small, withdrawal from Aandalsnes without loss of a single man. The official figures must show things up and there is bound to be a hell of a row.'

Straight was right about that. Parliament lost all confidence in the dithering Chamberlain. Labour members of the coalition government refused to serve under him. In dramatic circumstances Lord Halifax, an appeaser but the likely next choice as PM, withdrew from contention and the premiership passed to instead to Winston Churchill, the original mastermind behind the Norway venture, though he wasn't responsible for the delays that had alerted the Germans and doomed it.

So ended the 'Phoney war'. For his part in the abortive Norwegian campaign Straight was awarded the Military Cross, the Norwegian War Cross, and a grounding from flying duties. The doctors warned that he would be quite deaf for some time and made no promises that his hearing would ever be fully restored. So with a hole in either side of his head and another in his back he took the appointment of aide-de-camp to the Duke of Kent on the one condition, which he made crystal clear, that the very day he became fit for operations this would terminate. That day came earlier than expected, and when in December he passed the Central Medical Establishment's examination he was scolded for 'cavalier treatment of the Duke' in keeping his resolution by taking his leave that very day.

The Legion was at Exeter, commanded by Squadron Leader J. Anthony O'Neill DFC when Straight rejoined it, occupying comfortable flying club premises at the former civil aerodrome which had been requisitioned from

The founding father, Lord Edward Grosvenor. *(Author's private collection, from 601 Squadron's archives)*

Grosvenor and his embryo squadron. On his left (Air Commodore) Bill Langdon, Stores Officer, on his right (Air Commodore) George Bowen, Adjutant. *(Author's private collection, from 601 Squadron's archives)*

The squadron's Westland Wapiti light bombers, 1929–1933. *(The RAF Museum)*

Sir Philip Sassoon, Hon. Air Commodore and 601's second commanding officer, 1933. *(The RAF Museum)*

Hawker Harts in standard Vee ('Vic') formation. *(The RAF Museum)*

Porte Lympne, Sir Philip Sassoon's vast, Hollywood-baroque summer paradise in Kent. *(The RAF Museum)*

At Lympne aerodrome, from the left: T. E. Lawrence (of Arabia) in his mysterious assumed identity as Aircraftman Shaw, Sir Nigel Norman, Brian Thynne, Loel Guinness, Sir Philip Sassoon. *(Author's private collection, from 601 Squadron's archives)*

Trent Park, Sir Philip Sassoon's mansion north of London.

Sir Nigel Norman and squadron. Back, left to right: George Baker (Stores Officer), Lord Knebworth, Nigel Seely, Peter Clive, Reggie Elsmie (Ass. Adj.), Dr Crawthorne, Maurice Jackaman, Craven Hohler, Roger Bushell. Front: Rupert Belleville, Dickie Shaw, Dermot Boyle (Adj.), Sir Nigel Norman, CO, John Parkes, Brian Thynne, Loel Guinness. *(Author's private collection, from 601 Squadron's archives)*

Roger Bushell in 1937.

Rupert Belleville in the bull ring, Spain. *(Author's private collection, from 601 Squadron's archives)*

Hawker Demon that crashed on take-off. Guy Branch, the rear-seat passenger, won the British Empire Medal for Gallantry for pulling Aidan Crawley from the pilot's seat. *(Author's private collection, from 601 Squadron's archives)*

Roger Bushell, John Hawtrey, and Sir Nigel Norman. *(The RAF Museum)*

Willie Rhodes-Moorehouse and Amalia Demetriadi.

Clockwise from top left: Lord Grosvenor, Brian Thynne, Michael Peacock, Roger Bushell. *(Author's private collection, from 601 Squadron's archives)*

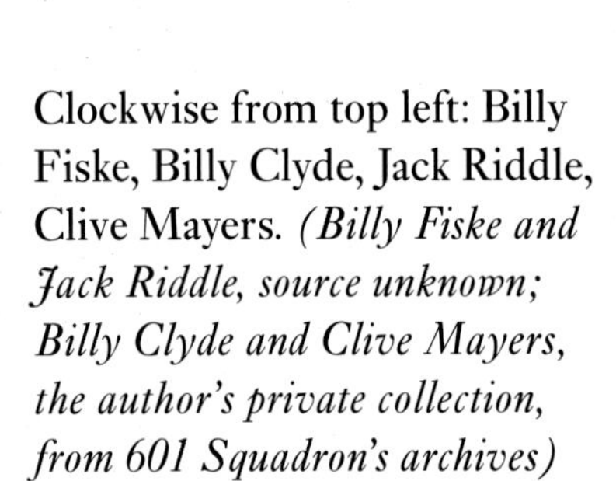

Clockwise from top left: Billy Fiske, Billy Clyde, Jack Riddle, Clive Mayers. *(Billy Fiske and Jack Riddle, source unknown; Billy Clyde and Clive Mayers, the author's private collection, from 601 Squadron's archives)*

At RNAS Ford with a Blenheim, summer camp 1939. From the left, with uncertain names in parentheses: Standing: ? ? ? Whitney Straight, Dick Demetriadi ? Huseph Riddle, Jack Riddle, 'Nono' Rathbone, Tom Waterlow the adjutant, (Tom Hubbard), Guy Branch, Willie Rhodes-Moorehouse, ? ? (Julian Smithers). Seated: ? Gerald Cuthbert, Charles Lee-Steer, Sir Archibald Hope, Henry Cavendish, (John Peel), Roger Bushell, Brian Thynne the CO, (Bob Foley), Michael Peacock, (James Little), Max Aitken, Raymond Davis, 'Paddy' Green, 'Doc' Williams. *(Author's private collection, from 601 Squadron's archives)*

Michael and Carol Doulton. *(Paul Doulton)*

Hawker Hurricanes with 601 Squadron UF markings. *(The Imperial War Museum)*

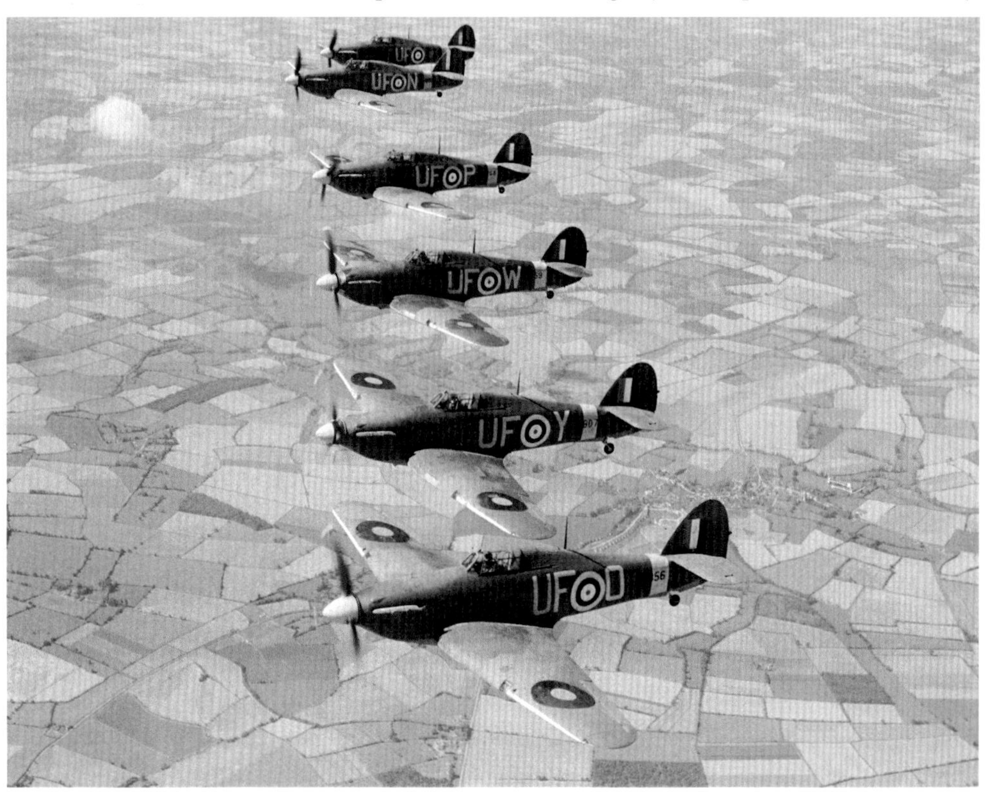

The crew room at Tangmere during the Battle of Britain. Right to left: Max Aitken, Dick Demetriadi, Charles Lee Steere, Willie Rhodes-Moorehouse. Within weeks all but Aitken would be dead. *(The RAF Museum)*

Right to left: Jerzy Jankowiecz (Poland), Archie Hope, Hugh Joseph ('Huseph') Riddle, unknown. Within days Jankoweicz would be killed over Dunkirk. *(Author's private collection, from 601 Squadron's archives)*

Willie Rhodes-Moorehouse. *(Jack Riddle)*

David Niven at The Ship. Opposite him sits Primula Susan Rollo, his future wife. The others are owners Trevor Moorehouse and wife Nanky. *(Author's private collection, from 601 Squadron's archives)*

Michael Robinson standing in the hole a Bf 109 canon shell made through his Hurricane's wing. *(Author's private collection, from 601 Squadron's archives)*

Max Aitken with his father Lord Beaverbrook, Minister of Aircraft Production.

Michael Robinson as wing commander, Tangmere. *(The Imperial War Museum)*

The grand arrival. The Associated Press announced, 'American Airacobras now fight with the RAF. They are doing a splendid job'. Nothing could have been less true. *(The Associated Press)*

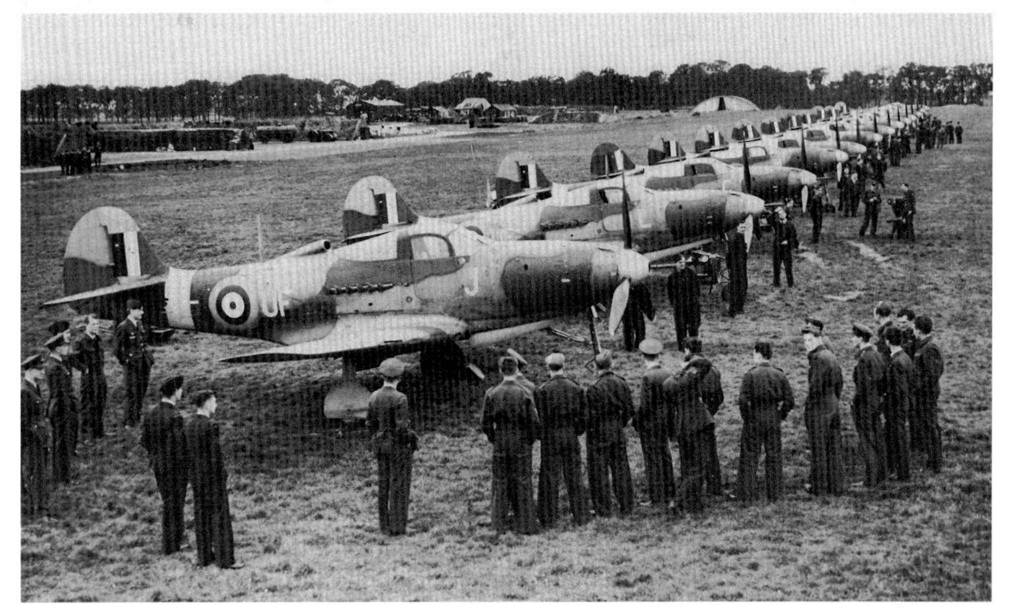

Squadron Leader Gracie exiting from an Airacobra.

No. 601 pilots at Exeter. Central is CO Squadron Leader J. A. O'Neill; on his right is Whitney Straight. *(Author's private collection, from 601 Squadron's archives)*

Malta hustle: refueling, rearming, and refitting a Spitfire.

A captured Stuka in the Western Desert, used against all regulations for utility work. *(Author's private collection, from 601 Squadron's archives)*

Spitfire 5Bs of 244 Wing over Tunis. The farthest two with 'UF' markings are from 601 Squadron. *(The Imperial War Museum)*

Flight Lieutenant Denis Barnham in a Spitfire, Malta, 1941. *(The Imperial War Museum)*

601 Squadron ground crew in Italy holding a Flying Sword. *(Author's private collection, from 601 Squadron's archives)*

Desmond Ibbotson as sergeant pilot. *(The Imperial War Museum)*

The squadron's last de Havilland Vampire. In the background, replacement Gloster Meteor 4s. *(Author's private collection, from 601 Squadron's archives)*

Seven of 601's commanding officers in the town headquarters after the war. From left to right: Paul Richey, Hugh 'Cocky' Dundas, Brian Thynne, Sir Archibald Hope, Sir Max Aitken, AOC Sir Sholto Douglas, R. G. 'Dickie' Shaw, Christopher McCarthy-Jones. *(The author's private collection, from 601 Squadron's archives)*

Squadron Meteor 8s over Valletta, Malta, during summer camp, 1953. *(IPC Media)*

The entire squadron at summer camp in Malta, 1953. Back row: Auxiliary and Regular ground crews. Middle row, left to right: John Merton, Harold Harmer, Ken Askins, Alf Button, Edward Goss, Nevil Leyton, Norman Nicholson, Teddy Lanser, Fred Triptree. Front row, left to right: Len Brett, Mark Norman, John Bryant, Jimmy Evans, Clive Axford, Emanuel Galitzine, Peter Vanneck, Peter Edelston, Christopher McCarthy-Jones, George Farley (Adj.), Dick Smerdon, Harry Davidson, Tom Moulson, Denis Shrosbree, Frank Winch, Jock Spence, Desmond Norman. *(IPC Media)*

The author, Tom Moulson, in Malta. *(The RAF Museum)*

The AOC, Earl of Bandon, astonishing the squadron with a tour de force in Malta. *(Author's private collection, from 601 Squadron's archives)*

Prince Philip, Duke of Edinburgh, the squadron's last honorary air commodore, takes the salute at its last march past. *(The Zuma Press)*

A 601 Spitfire relic of 1941 in Malta, seen in 1954. *(Author's personal collection)*

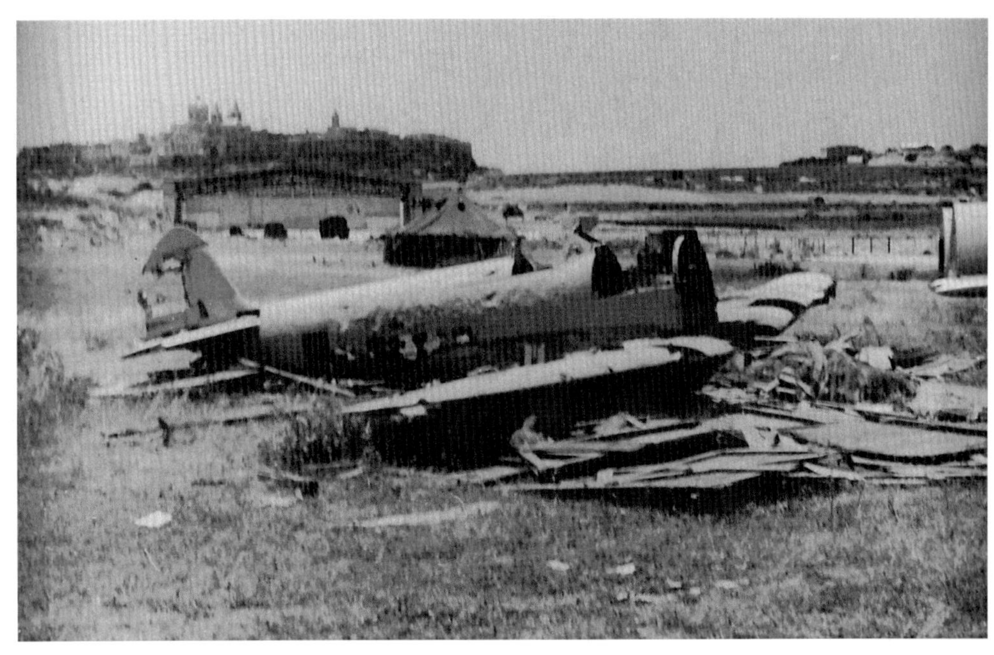

Past members of the squadron parading, with Archie Hope front row right and Dickie Shaw next to him.

Jack Riddle at the dedication of a specially designed window commemorating Billy Fiske at Boxgrove Priory, site of his long-neglected grave, with a replica of Fiske's Bentley. *(Ed McMahon)*

Ray Sherk of 601, right, and Horst Wilborn, the man he shot down over the Western Desert in 1942, in Hamburg nearly seventy years later. *(Ray Sherk)*

The photo that lives on! More than sixty years after the Battle of Britain this 1940 press photo of No. 601 Squadron pilots scrambling at Exeter, sponsored by the Air Ministry, remains a favourite of books and magazines worldwide. In fact the action was staged, and looks it; but the photo survives as a compliment to 601.

its owner, a certain Mr Whitney Straight. Once more a flying officer, he celebrated his return within a week by destroying an He 111 on 12 December 1940. Having missed the Battle of Britain, this was his first victory.

Increasingly 601 was engaged in shipping patrols and bomber-escort missions over France and Belgium. Straight was promoted to 'B' Flight Commander and led twelve Hurricanes in support of Blenheims from No. 139 Squadron in a raid on Boulogne harbour. On the return across the Channel he saw a black, yellow nosed Me 109 below flying in the opposite direction, and after ordering his section to continue Straight made a diving turn onto the unaware enemy aircraft, which after one burst spun into the sea.

There were frequent freelance offensive sweeps, called 'Rhubarbs', over German-occupied Europe, in which the purpose was to devastate more or less randomly motor transport, canal barges, airfields, and any other useful target that presented itself. Less offensive missions involved dropping propaganda leaflets through the flare chute upon enemy-held towns, and on such a flight Straight circled the Belgian town of Ghent at rooftop height with his number two, then flew down the main street at upper storey level to the acclaim of civilians who waved from the windows or ran out on to the pavements, preventing, as it happened, the Hurricanes from shooting at a cluster of German troops who stood at the end of the street. Returning at low level across the Channel, the two fighters audaciously attacked a Nordfleet flak ship, raking it with bullets just below the funnel.

The flak ship, whose sole purpose was to shoot down RAF aircraft with its formidable battery of guns, was hardly a favourite or indeed a wise choice of target for fighter pilots, but Straight was lucky. He received from now-AVM Sir Trafford Leigh-Mallory 'another nasty job with no future' as he described it – command of No. 242 Hurricane Squadron, escorting Blenheim bombers seeking targets in the Channel. On 31 July 1941 he attacked a flak ship between Le Havre and Fecamp. The flak ship opened up with everything and a shell hit his engine. Straight force-landed in a field near the coast.

Dazed, his forehead and knee cut and bruised from impact with the instrument panel, he attempted unsuccessfully to fire his machine with matches borrowed from a wide-eyed worker in a nearby field before running two miles and then settling down to a fast walking pace. He had always been prepared for a landing in enemy territory and he checked his possessions while walking: RAF uniform, the specially designed suede leather golfing jacket into which were sewn three passport photographs and 700 francs in notes, service ankle boots which might pass as civilian shoes, and a loaded automatic pistol. His intangible but invaluable assets were fluency in French and a knowledge of France's geography. It was imperative to move fast (ten minutes after the crash a Frenchman notified the occupation authorities) and this involved taking some risks.

He hurriedly buried his uniform jacket and bought a plain civilian cap for three francs at a farm house. There were few people and no soldiers on the road. Within an hour he found a railway station with a train, bound for the south, standing at the platform. He crawled up the embankment and into the last coach without being seen, and was just about to relax when the train pulled out, leaving his coach behind.

Straight set off again on foot, and by curfew time had reached a point three kilometres from Belbec. He was exhausted, and his exhaustion drove away caution. Knocking at a farm door, he asked the occupant for shelter in the farm, and then went on to explain his case frankly. The farmer, terrified, refused, and said there were German troops using his barn as a billet, and slammed the door. Deciding he badly needed rest to stay alert, Straight slept the night in the barn under a pile of straw, undisturbed by the occasional coming and going of German soldiers. Almost in his sleep Straight weighed the possible escape routes he had mentally rehearsed a hundred times. He could make contact with friends in Marseilles or Biarritz or cut through the unoccupied zone of France to neutral Spain. But first there was an all important call at the American Embassy in Paris where personal friends could give him a haven, a hot bath, money, a good supper and a fast car to his chosen route.

At dawn he left the barn and walked to the small railway station at Belbec where he bought a ticket to Paris without difficulty. At Rouen, between trains, he took the opportunity of getting rid of his English coins by distributing them in the letter boxes of German billets.

That evening he was in Paris. Only after trying several hotels was he allowed to stay the night without papers. In the morning he spruced up, adjusted his cap French-wise and set off for the American Embassy. Here he received the first shock since his forced landing; a notice on the door announced: 'CLOSED. ALL ENQUIRIES SHOULD BE ADDRESSED TO THE UNITED STATES EMBASSY, BERLIN.' Stymied, he rang the doorbell. The caretaker, an American, appeared. As Straight wasted no time developing his story and outlining his needs, the man, distressed, began to cry. 'Sure, Mr Straight,' he sniffed, 'we heard about you from the BBC and now the Germans are looking for you everywhere. See that building across the street? That's the military headquarters and they can see you from the window if they look. Please, Mr Straight, leave us alone.'

The door closed and Straight walked briskly out of sight of the embassy building. He was shaken by the elimination of his greatest escape hope. How the BBC could have been so mindless was beyond his understanding. The Germans, always relentless in their search for airmen who had been shot down, were never more so than when a 'name' was involved. A few minutes earlier he had thought his greatest asset was to be Whitney Straight; now it was his greatest liability. Two German soldiers were approaching along the pavement, and he turned quickly into a café.

Over a coffee he began to revise his evasion plan. Already he couldn't fail to notice the intensity of anti-British propaganda – in the press, on the wireless, in the sullenness and fear with which everyone to whom he turned for help treated him. It would be better to make for unoccupied France where surveillance would be slacker and the people more approachable, and then to head south across the Spanish frontier. But he would need far more money than he had.

There was a telephone in the corner of the cafe and he put a call through to the American Embassy. He recognised the caretaker's voice. 'Please listen to me,' said Straight, 'I'm round the corner in the Café X. My only desire is to get away from here as quickly as possible. I don't want to cause you any trouble. If you will just send round some money I can carry on alone, and I promise that you will hear no more of me.' It sounded like blackmail, and perhaps the caretaker thought so too, for he said he would see what could be done and rather to Straight's surprise the money was delivered. A Frenchman entered the café, walked immediately up to him, pressed 12,000 francs into his hand, and departed without saying a word. The business took ten seconds.

Straight bought a Michelin road map, which helpfully showed the demarcation line between occupied and unoccupied France, and took the next train to Tours. The greatest problem was his lack of food coupons, almost as valuable as money, and a constant diet of ice cream, sweets and coffee was already beginning to deplete his stamina. The Michelin map showed a station at Chenonceau, near Tours, only five hundred yards from the River Cher, which served as the boundary. Here he could cross into Vichy France.

At four o'clock in the afternoon his train puffed into Chenonceau on the River Loire and he walked the short distance from the station to the river bank. A few anglers sat scattered around peacefully on the near side, and a German patrol marched past at intervals. After timing the patrol intervals at fifteen minutes, Straight waited for it to pass a fourth time and then walked down to the edge of the river and lay down on the bank near one of the anglers. He realised for the first time a fact that the Michelin guide hadn't revealed: he was in full view of the huge fifteenth century Chateau Chenonceau, which he could see from the swastikas flying over it was the local German military headquarters.

The patrol returned, and after it had passed Straight jumped fully clothed into the water and began swimming. The current was strong and carried him nearly down to the chateau before he dragged himself clear among the reeds on the opposite bank. The anglers paid no attention to him, so he took off his clothes, wrung them out and sunbathed for about an hour as they dried. Then revived and exultant, he walked to the nearest village and obtained an excellent dinner and accommodation for the night. Though he was unable to produce an identity card and his clothes were still damp, no questions were asked.

By the evening of the following day he had reached Toulouse by bus and train, and he bought a map of the Spanish frontier. Another night in a hotel and another train saw him in Pau, only thirty miles from the border. At Pau he changed trains and continued towards the Pyrenees, now clearly visible and temptingly close. At Bedous a gendarme boarded the train and demanded to see all identity cards. With a great show of concern Straight 'discovered' he had left his behind at Pau and said he would return for it immediately. The gendarme accepted this explanation – it wasn't too uncommon an occurrence – but Straight had to leave the train. He spent the rest of the day on foot, and by nightfall was within sight of the border. It was an exciting moment. But he was too tired to continue and his feet were in bad shape, so deciding that a night's rest would sharpen his wits for the border crossing he booked a room in a small house.

The village was teeming with French troops, but he could only find one café and although it was busy he entered and asked for a meal, explaining that he had no meal tickets but that he would pay double. The proprietor accepted. It was the kind of risk Straight had taken before. A civilian at the table where Straight took his seat finished his meal and left within minutes. Conscious of his achievement and the proximity of his freedom, Straight experienced a familiar wave of elation but at the same time a tinge of apprehension. With only a day to go, it wouldn't do to be incautious. He wouldn't eat again until clear of Vichy France.

As he left the café a figure stepped out of the shadows and barred his way. It was the civilian, now wearing the uniform of a gendarme. He asked to see Straight's identity card and papers. Straight replied that they were in his room at the house, and they walked there together. He went up to his room and promptly jumped out of the rear window, but the gendarme was waiting there with a drawn pistol and took him prisoner as his feet touched the ground.

After a night without food or water in a cell at Bedous, Straight realised his position was hopeless, and he admitted being a British officer – not Whitney Straight because he knew RAF prisoners had a reputation for attempting escape, suffered greater restrictions, and were more likely to end up in a German Stalag Luft than was a run-of-the-mill army officer. He became Captain John Willard of the Royal Army Service Corps. His captors immediately warmed in their attitude and gave him food and cigarettes. He was interrogated in the most cordial way by the Colonel of the *Département* at Oleron, who gave him ration tickets, parole, a room in a hotel, and a sincere apology for having to place any restrictions on him. When he was eventually conveyed to the internment camp at St Hippolyte, Straight decided to stay out of trouble for a while, recover his strength and discover the best way to escape.

The atmosphere at St Hippolyte was a strange one for the potential escaper. The guards were friendly but dedicated to preventing escapes; their families lived in occupied France and the Germans had those addresses. Straight's first attempt at escape was very crude. He met an old air force acquaintance, Squadron Leader Gibbs, and with a second lieutenant named Parkinson they planned to rush the front door of the camp, Gibbs first, followed by Straight and then Parkinson. The guards grabbed Gibbs's coat as he ran through the gate, but he managed to wriggle out of it and, with Straight and Parkinson harassing the Frenchmen, escaped to a nearby café where contacts were waiting with a bicycle. But the guards had drawn their revolvers, and the other two gave up their attempt. Straight was even able to bully the camp commandant into believing they hadn't tried to get out.

Straight's next plan was subtler and more ambitious. There were about three hundred internees at St Hippolyte, a mixture of nationalities but mostly the British remnants of the Highland Division, captured at the time of the Dunkirk evacuation. Increasingly RAF aircrew began to arrive, shot down on sweeps over France, and it became clear that there were some good aircrew worth sending back to Britain. For some time, those prisoners declared unfit for active service had been repatriated at irregular intervals on a certificate from the medical authorities. Straight set his heart upon the wholesale repatriation of the aircrew, encouraging them to fake medical evidence and teaching them every device he knew. The most popular method of faking disease was to swallow a rolled cigarette paper which showed a realistic ulcer under X-ray. But if some of the internees had trouble with these artifices,

Straight had little. The wound on his back was genuine, and lumps of butter in his ears converted his scars into fearful running sores. A liberal intake of aspirin before each examination threw his heart out of rhythm and lent him a convincing air of stupor. He was also a reasonable actor. The ear specialist at Nimes was completely taken in and issued him with a certificate with which he secured the necessary documentation from the camp's military doctor. Forty-two other internees, including several of those whom Straight had prepared for repatriation, received similar documentation.

It was clear that the French were being as helpful as they could. A French colonel even gave Straight information on German installations and troop movements in the north, which he coded on a scrap of paper and concealed in the turban of an Indian Army Officer due for repatriation. There were considerable delays, but on the assurance of the American negotiators that everything would soon be concluded Straight made no further attempt to escape. On 15 January 1942, six months after his first capture, he learned that a letter signed by Admiral Darlan had been received by the camp administration, authorising repatriation of the forty-three internees. On 5 March, only two days before expiry of the Spanish visas, the party left St Hippolyte. At Perpignan it was turned back on orders of the Vichy government, and next day was back at St Hippolyte.

The disappointment was intense, and Straight probed to find out what had changed the government's mind. He learned that on the night of 3–4 March, thirty-six hours before the departure from St Hippolyte, the RAF had bombed the Renault factory at Billancourt, near Paris, with 325 aircraft. Although he didn't know how many Frenchmen had been killed [391 plus 558 seriously injured], he sent a message to Mr Anthony Eden in London, deploring 'the killing of the innocent people of Paris', the sentiment of this message being intended for the French intermediaries who undertook its delivery.

As the weeks passed he began to suspect dilatory handling of the paper work by the Americans. Several times he let it be known to the Resistance that he was contemplating escape but time and again was assured that the repatriation would go through, and that any escape bid on his part would only prejudice the others' chances. This posed a painful dilemma. As the months went by Straight was being prevented from taking action by a promise, the worth of which seemed steadily to diminish. His faith in repatriation was wearing thin. It wouldn't have been impossible to accept life at St Hippolyte and await events, but if he couldn't get back to England one way the Straight was determined to do so otherwise. When on 17 March the party was moved to a new internment camp at Fort de la Revere and he saw the formidable character of his new prison, Straight set 15 May as the date beyond which he could place no further reliance on the chances of repatriation.

The set day dawned and passed and he made up his mind. The fort was difficult for escape, so he had himself removed to hospital by complaining of headaches and ear troubles, by this time not entirely untrue. At the hospital the system of guarding had been tightened up, although it was slacker than at the fort; windows were permanently barred and the doors locked each night; sentries patrolled the outside twenty-four hours a day. It was essential to make the escape by day when vigilance was less strict and the doors were unlocked. He joined a Polish sergeant and an English corporal for a joint effort, passing the word to the Resistance movement through a tradesman. They were on the sixth floor, and their only exit was the main door at ground level. They told the Resistance of this problem, and the tradesman delivered a phial of drug pellets. There were no dosage instructions and the last thing they wanted to do was to kill the guard, so Straight's choice of 'three for luck' was purely a guess. As Straight was due to be sent back to the camp on 23 June, they planned the break for the 22nd. At lunchtime that day they diverted the guard's attention and dropped the pellets in his wine. Within minutes the guard was asleep. They changed their pyjamas for civilian clothes, but the guard regained consciousness and began walking unsteadily downstairs. Wishing they had made the dose one pellet stronger they followed down the six flights of stairs at a safe distance. When the guard turned left at the foot of the stairs they turned right and bolted, missing by luck the second inside guard. Everything happened so swiftly that the four gendarmes on duty outside were easily given the slip.

There was uproar at the camp when news of the escape broke, and guards who were able to recognise them were posted at every railway station in the area. But Straight's incredible luck didn't desert him. Helped by the Resistance, which supplied papers, forged identity cards and a guide, he succeeded under the guise of 'Charles Dumontel, Engineer', in passing several of the guards whom he recognised at the check points. He had a moment of uneasiness while sitting in a bus opposite a woman he had known in Paris before the war, but to his relief – and minor disappointment – she showed no sign of recognition.

At 4 a.m. at an appointed place on the deserted beach near Perpignan, a rowing boat took Straight and several other fugitives out to a jaunty, innocent-looking trawler flying Portuguese colours lying a mile offshore. Some days later, after being repainted overnight, the same ship sailed into Gibraltar, her hull crammed with machine-guns, resplendent in the colours of the Royal Navy.

Chapter 8

Wind the Windows Right Down

The reason birds can fly and we can't is simply that they have perfect faith.

John Matthew Barrie, *George Meredith*

At Exeter the Legion almost lost its soul. The Class 'C' squadron was a creation of Dowding's as an alternative to the completely non-operational unit. Only a quarter of the operational pilots on 601 were retained, still led by Archie Hope. The posting away of senior pilots and influx of new, inexperienced ones tore away at the fabric of squadron cohesion and *esprit de corps*.

It has been suggested that the intimacy of an Auxiliary squadron, especially one with family relationships, makes it more vulnerable to a collapse in morale than that of a regular unit with its members' icy detachment. The Legion's experience proved the contrary. Personal loyalties had pulled the pilots closer together and strengthened their resolve. That went to the essence of *esprit de corps*, trumping concern for self and even for country. It was the erosion of this almost tribal unity that cost 601 its loss in morale.

There was action, but no daily sharing of common peril against the enemy, little contact with the Luftwaffe. The worst symptom of low morale was an appalling accident rate, Hurricanes crashing on take-off, turning over on landing, and colliding in the air. Michael Robinson was now in charge of training and tried to 'inculcate the fight-pilot spirit', as he described it. His eyebrows would meet in a scowl as he smacked his fist into his palm and explained his technique to new arrivals. 'Get into an aircraft,' he would order, 'any aircraft; leave the throttle open, turn round the church spire at naught feet and be back here in five minutes!'

Hope wasn't sure Robinson's method would reduce the accident rate and had to call it off. But he shared Robinson's irritation when a national daily reported that a voluntary London group had sent gifts ('comforts') to every one of the capital's armed service units, which apparently excluded 601. A cool but courteous letter to the newspaper attracted a welcome shower of woollen mufflers, mittens, balaclavas, food hampers and, as it was nearly winter, a Christmas card.

Robinson was to move to other squadrons and accumulate nine kills. On 10 April 1942, while leading the Tangmere Wing, Wing Commander Michael Lister Robinson, DFC DSO and Belgian Croix de Guerre, failed to return from a sweep. Neither he nor his aircraft was seen again.

The Air Ministry tried to bolster squadron morale by sending a war artist to paint portraits of its aces. One wonders where the ministry got its ideas from. Since picking aces would had made matters worse, Hope had the artist paint a composite picture of all the pilots assembled round a Hurricane.

On 18 December, 601 moved to Northolt to resume operations under Squadron Leader O'Neill. It was an opportunity for the squadron to regain its self respect, though a slender one, for there was a change of rôle. With Germany's abandonment of mass daylight raids, a new RAF policy vaguely defined as 'leaning forward into France' came into effect. Germany was now aware that the war would have to continue with Britain still in it. Not too concerned about this though dismayed by the dramatic failure of his air force, Hitler shelved the invasion plan indefinitely and turned his attention to the east, to Greece and Yugoslavia.

Fighter Command struck at trans-Channel objectives, marauding German-dominated Europe and attempting to attract the Luftwaffe to the western periphery and relieve the pressure on Russia. The missions 601 now undertook from its new base at Manston, on the eastern extremity of Kent, were aggressive, provocative, variegated, and dangerous. It was on such a mission that Whitney Straight would be shot down. The 'Rhubarbs' (fighter-bomber attacks) and 'Circuses', short-range flights across the Channel, descending through cloud to attack targets of opportunity, both intended to keep the German fighters engaged, involved heavy casualties. It was falsely supposed that the enemy's losses were even higher – precisely the mistake Goering had made during the reverse situation when he softened up British defence in July 1940.

O'Neill was good for the squadron and raised its morale. An entry in his log book in March 1941 says it all: 'Sqdn airborne in 2½ minutes.' When he was shot into the sea off Dungeness and taken to hospital with leg wounds he was replaced by Squadron Leader E. J. Gracie, DFC, a thick set and fearless fighter leader whose respect for his 'boys' matched theirs for his flying skill and dauntless leadership. A ground officer who had remained a pilot officer for some time and suggested to Gracie it was time to be promoted was told,

'My boys come to this squadron as pilot officers and die as pilot officers. Consider yourself bloody lucky if you manage to remain one until the end of the war.'

The air war for the Hurricane was petering out. Having made its debut over France and won its laurels over England, it was due for new roles in the Middle East. It was a safe guess that new aircraft would be found for 601. On 1 July1941 the squadron moved to Matlask where the Hurricanes were gradually taken away and replaced with a new fighter from America, the Bell Airacobra. Attempts were now made to bring the squadron up to the operational level on the new machines. These attempts were doomed to failure.

For seven months a hopeless battle was waged against the Airacobra both in the air and on the ground. The contest ranged over three airfields. At Matlask the combatants met, at Duxford the tussle began, and at Acaster Malbis, aided by bad weather and a bad airfield, the Airacobra won hands down. Designed and built by the Bell Aircraft Corporation in Buffalo, New York to a US Army Air Force specification, 240 Airacobras had been ordered from the manufacturers by Britain. A delivery to Southampton was uncrated and assembled at the nearby RAF base of Eastleigh, and three Airacobras were sent to the Fighter Development Unit at Duxford for testing. The results were dismal. The take-off roll was unacceptably long, and one test pilot disappeared over the hedge and was thought to have crashed, but managed to hold his height long enough to build up climbing speed. Above fifteen thousand feet, where most activity in Britain and Europe occurred, its performance fell off drastically owing to lack of a turbo-supercharger.

The first two aircraft were delivered to 601 at Matlask on 6 August. All the pilots were keen to receive new machines, but the new arrivals immediately prompted doubts. A good aircraft usually looks good, and experienced pilots can often judge a plane's handling by sight. This one had a highly unorthodox design and didn't look at all right. It was single-engined with its Allison V-12 motor behind the pilot's seat and a ten foot propeller shaft running under his seat. It had a nose wheel instead of a conventional tail wheel for improved forward visibility on the ground and to house a mammoth 37 mm cannon, roughly twice the diameter of an Me 109's cannon shells, through the propeller hub. It was said to be the first fighter designed around its armament. Though its weight was about the same as that of the Hurricane and Spitfire, its wing loading was one fifth higher, meaning its take-off and landing speeds would need to be greater. The central placement of the engine was intended to give better manoeuvrability but led to doubts about the longitudinal stability of the aircraft. The ground crews, for their part, took one look under the various panels and through the compendious servicing manuals, and predicted trouble.

Entry to the cockpit was via a wide hinged door, not a conventional sliding hood. The door had windows that could be raised and lowered as in a car, but the door assembly's construction was flimsy and the Pilot's Notes warned, 'It is THEREFORE ESSENTIAL TO WIND THE WINDOWS RIGHT DOWN before slamming the doors shut or the windows may be broken by cracking.' The undercarriage and flaps were operated by either an electric switch or manually; if set to normal, that's electric operation, the winding handle would be rotated by the motor causing serious risk of injury to the pilot. *Aeroplane* magazine reported that 'The Airacobra has all the disadvantages of both pusher and tractor (propeller) layouts, with some more of its own.' The Pilot's Notes cautioned, 'At idling speed the whole aeroplane shakes violently. As a result the engine should not be run on the ground at less than 1,000 rpm except when landing, to keep the landing run within reasonable length, and momentarily when necessary for manoeuvring. Roughness is also evident in flight but seems less at about 1,800 rpm.' Pilots of the Air Transport Auxiliary (ATA), who delivered aircraft to operating airfields, were advised to fly the Airacobra at whatever rpm caused the least vibration. The ATA pilots, mainly women, flew aircraft without radio and often without blind flying instruments. Such instruments as they sometimes had were only a gyro turn and bank indicator in addition to airspeed and altimeter, but no artificial horizon. With these only a skilled pilot might maintain safe flight in cloud. A year earlier Billy Clyde and Max Aitken had met an ATA pilot in the One Hundred Club in London and were so horrified to learn that she'd received no such instrument training they spent hours explaining it to her with napkin drawings and insisting she never enter cloud until she had practised sufficiently. The Airacobra wasn't ideal for the inexperienced. A woman ATA pilot delivering an Airacobra from Kirkbride to Hawarden was killed when it exploded and crashed on take-off owing to excessive rpm.

Another concern was the unreliability of the nose wheel 'down' light, meaning a pilot might not know whether the wheel was down, and there were fears that if it wasn't the engine would tear forward off its mountings and vulcanise the pilot against the nose structure.

On 8 November the Under Secretary of State for Air, Lord Sherwood, accompanied by the American ambassador, Mr J. J. Winant, visited Duxford and inspected the planes, their servicing crews and pilots. Air Ministry and press photographers swarmed around the Airacobras to cover this important political event, the first delivery of American fighters to an RAF squadron. There was much goodwill and the visitors reflected none of the squadron's gloom, leaving apparently well pleased with what they saw of the formation fly-past, practically a total effort. There were numerous press photographers present for this important hands-across-the-sea event. The Associated Press had no reservations and exulted, 'American Airacobras now fight with the RAF in Britain. They are doing a splendid job. Mechanics of the ground staff give the Airacobras a thorough overhaul to keep the planes in trim fighting order.' In the afternoon of the same day the squadron was privileged with a similar visit by the Russian Mission, and the hopeful rumour was circulated that the machines would be sent to Russia. (They were.)

Sergeant Briggs flew the first Airacobra eight days after its arrival. While attempting aerobatics the aircraft's first vice was confirmed; it tightened up in a turn and blacked him out, and when he recovered Briggs was so dizzy he diverted to Mildenhall, where a fault in the electric circuit jammed his undercarriage and he had to crash-land. On the same day, Sergeant Bell, namesake of the Airacobra's designers, force-landed the second machine at Oakington after a glycol leak slowed the engine revolutions.

Two days later the squadron transferred to Duxford, more Airacobras arrived, and familiarisation proceeded in earnest. Flight Lieutenant Himr force-landed at Marham when his driving shaft fractured. Sergeant Land crashed at Langham when his fuel supply failed, and was taken to hospital with concussion. Sergeant O'Brien crash-landed in a field near Oakington but was uninjured.

Four planes were detached to Manston for trial in actual operations. Though fog curtailed activities, they strafed enemy troops on Dunkirk pier and along the beach popularly known as Blackpool Front, and fired on canal barges. After six days and nothing more conclusive the four pilots returned to Duxford and drew a blank on their experiment.

Within a few days Pilot Officer Hewitt, while practising aerobatics in the Bedford area, for no known reason spun into the ground and was killed. Sergeant Land's engine cut on take-off and he crashed through the boundary hedge. Pilot Officer Sewell's fuel supply failed while he was in the circuit but he was able to land on the airfield. Sergeant Sawyer had the same trouble but was killed when he crashed near Debden.

On Christmas Eve of 1941 Gracie left for what was to be the toughest assignment of his career. Nobody realised that he was shortly to meet the squadron again in vastly different circumstances; morale was at a low ebb, and the future defied imagination. When in the New Year the squadron moved to Acaster Malbis things took another turn for the worse. The Airacobras resisted most efforts to make them serviceable while snow and slush made the airfield impossible for take-offs and landings. During the five days from 20 to 24 January the field was completely hopeless, and a single flight to Church Fenton on 25 January registered only ten minutes' flying. For the last five days of the month two weather tests clocked up one hour and twenty-five minutes. Total squadron flying for the month amounted to twenty-two hours and fifty minutes, considerably less than a morning's effort during the Battle of Britain.

The progress report, a depressing document to compile, transmitted some chill from the frozen meadow that was Acaster Malbis to the lofty corridors at group headquarters. To see for himself the nature of these difficulties, AVM ('Birdie') Saul CB DFC came to stay the night. With so little flying to disturb them the ground crews had managed to make each of the thirteen Airacobras serviceable but the field was blanketed with snow and the crew and pilots spent all day ploughing and shovelling it away. Even if the prime minister and the entire Air Council had come 601 could not have pretended to be airworthy.

When persistent effort had prepared a runway and the weather made flying possible, Pilot Officer Kenneth Pawson resolved with a mind for realities to test the Airacobra as a fighter. He climbed to twenty-seven thousand feet over the sea to try out the guns and in the usual atmospheric temperature of minus fifty-six degrees Fahrenheit first flew around for twenty minutes to stabilise the aircraft's temperature, then fired all guns simultaneously. The central cannon fired ten rounds before its magazine froze and one of the port .5s failed completely with a frozen breechblock.

On 12 February the fact that four of the serviceable six aircraft were airborne simultaneously was considered worthy of celebration. Next day Pilot Officer McDonnell attempted an ordinary roll at six thousand feet within sight of the airfield. While on his back he appeared to lose control, his tail dropped and a cloud of smoke from the exhaust showed that he was trying to correct the longitudinal trim by revving the engine. His aircraft then fell, still inverted, in a steepening dive and crashed on the muddy bank of the Ouse.

On 10 March 1942 Squadron Leader John B. Bisdee DFC arrived to take command of the squadron. He was a pre-war VR sergeant pilot who had acquired his operational experience with one of 601's northern relatives, No. 609 (West Riding of Yorkshire) Squadron. Besides being a man of will and determination, a bulldog type with Grosvenor's physique, he benefited greatly by arriving fresh on the scene and brought with him a crisp vitality and optimism that would never have been hard to generate from within the unit. Bisdee found himself at the head of a body of men who sensed that they had lost their way and were pulling at a tangent to the war effort. What had formerly been a crack fighter squadron now felt that it was being treated as a Cinderella, given aircraft that couldn't be flown, classed as non-operational and left to flounder in a Yorkshire quagmire.

The Airacobra was a broken reed, unpopular and unsuccessful. Even if the machines were made serviceable by the morning they would almost invariably go 'U/S' on the first engine run-up. There were many electrical troubles from a system that was better suited to a dry North American climate than to the cold dampness of an English winter. When occasionally one took the air, it would go popping and banging around the sky, its pilot tensely waiting for the next thing to go wrong. The ground crews, always good barometers of squadron morale, were depressed and had arrived at the unfortunate point of expecting servicing troubles. Though the ground crew contained many pre-war Auxiliaries, only one remained among the pilots, and he soon left. A shining example of the quality of the squadron's pre-war ground crews, Flight Lieutenant Jenson had been a metal rigger on 601 in 1933, became a sergeant pilot during the Battle of Britain, and would finish the war as CO of a Typhoon squadron before going on to be a group captain.

Realising that the best way to pull the squadron together was to show it results for its labour, Bisdee realised that first he had to get it into the air in some way. Acaster Malbis, little if at all higher than the nearby Ouse and described by the pilots as a God-forsaken cow pasture, offered no promise, for even as the spring sunshine began to melt the snow the heavy Airacobras persistently sank their small nose wheels into the soft mud. Either the aerodrome or the Airacobra, Bisdee realised, would have to be changed. First he tried to use neighbouring Church Fenton, which had runways, but there too many Blenheims and Beaufighters choked the circuit with take-offs and landings. There was nothing for it but to pay a visit to the AOC.

'I understand your problem,' sympathised AVM Saul, AOC No. 13 Group, after hearing him out. 'You need a change of aircraft. Well, Bisdee, you'll be happy to know that it's already intended to equip you with Typhoons.'

This was an awkward moment for Bisdee. He had heard about Typhoons what everybody had heard. Their engines were failing and their tails were breaking off in high-speed dives. He didn't want to appear a whiner, but swapping one problem aircraft for another wouldn't help.

He paused, and tried to look as though he was carefully weighing the offer, then said, 'Thank you, sir. That would certainly be an improvement. But I'm not too sure it will do the trick. You see, my pilots are brassed off from lack of action. The ground crews are expecting servicing snags. I want the squadron to get its hand in as soon as possible and win back its self respect. I know the Typhoon's still having teething troubles. Could we possibly have an aircraft that's graduated – something that would be a shot in the arm for all my squadron?'

Saul looked at Bisbee and nodded slowly. They could both be thinking of only one aircraft. With the best of grace he said, 'Very well, Bisdee, I think perhaps that can be arranged.'

Thus did 601 get Spitfires, self respect, operational listing, and a new assignment second to none given a fighter squadron at any time, anywhere. The Spitfire was the best known fighter of its day. A descendent of the Schneider Trophy seaplane winner, with its elliptical wings and harmonious lines it was the most graceful and lusted-after Allied fighter of the war. Though its long engine nacelle made taxiing difficult and the narrow wheel track called for care during take-off, it was extremely easy to handle once in the air, light and responsive; the word often used to describe its handling was 'sweet'. It was the standard British fighter of the war, its name widely used as a generic for an RAF fighter. It was chopped, changed, re-modified, given a four-bladed propeller (the earliest ones had only two, later three) and thrown into a hundred different rôles. The squadron knew it was getting the best. Converting from the Airacobra to the Spitfire was like changing a sick Clydesdale for a thoroughbred. The squadron would never again fly any other type of propeller aircraft.

The unloved Airacobra was withdrawn from service, its remaining numbers crated again and sent to the Russian, Australian, and Free French air forces. After this, all future shipments were diverted to Russia. During its time with No. 601 Squadron five Airacobras were lost in accidents, all, of course 'due to pilot error'. Squadron morale soared. The undignified arrival of the first three Spitfires, when the one flown by Bisdee tipped on its nose into the mud, the other two abandoning their attempt to land and returning to the maintenance unit, discouraged nobody. Bisdee simply put in a demand for six hundred yards of Sommerfeld tracking, and when it arrived had the entire station complement, with the exception of cookhouse and kitchen staffs, remove their jackets and manhandle the wire mesh into place on the soggy ground. The Spitfires were then flown in, refuelled, towed into the Sommerfeld runway and immediately taken off for formation practice.

The time for tactical training soon arrived, and the Spitfires were flown in creditable formation to Digby, the remainder of the unit following by rail. For six weeks they practised formation, gunnery and dogfighting. Bisdee introduced his four-man snake, which had proved successful with No. 609 Squadron, four aircraft weaving from side to side through half circles in line astern so that at any moment every patch of sky was being covered by at least one pair of eyes. For once, the squadron had something of its own. (The advantage of the Luftwaffe's *schwarm* was still not sufficiently understood.)

Now that 601 was again a flying unit, Bisdee strove to make good certain deficiencies on the social side. He sent for the Intelligence Officer, Pilot Officer Kenneth Carew-Gibbs.

'Gibbie,' he boomed, 'What would you say was most lacking in the amenities of civilised life at Digby?'

'Off duty recreation, I should think sir,' replied Carew-Gibbs without difficulty.

'Spot on, Gibbie! The poor chaps have even forgotten what a civilian looks like. Something must be done about the deplorable absence of strongly forged links with the local gentry. I appoint you, Gibbie, to the post of Social Contacts Officer. See if you can pull a few parties out of the hat for the troops, and some influential Yorkshire manor houses with presentable daughters for my pilots!' It was typical of Bisdee that he prescribed so exactly the elixir of squadron life, and that he was able to mix the ingredients in their right order. It was typical of war that the cup was struck from the squadron's hand before it could taste the brew.

Bisdee took the telephone call in the officers' mess. Group must have awaited word that the famous No. 601 Squadron was at last operational as an impecunious heir awaits a legacy, and it wasted no time in presenting its claim. As the line wasn't scrambled, the voice from Group was guarded.

'Squadron Leader Bisdee, we have glad tidings for you. You will be delighted to know that your squadron is being posted overseas.'

'I wonder where?' Bisdee ventured.

'Ah, we can't tell you that officially for a couple of days,' the cheerful voice from Group headquarters continued, 'but I can personally assure you that it won't be more than a thousand miles from a little island in the middle of the Mediterranean.'

Chapter 9

The Crucible of Malta

Oh, the Battle of Brit all over again!

Squadron Leader Bisdee, within minutes of landing in Malta

It all makes the Battle of Britain and the fighter sweeps seem like child's play in comparison, but it is certainly history in the making, and nowhere is there aerial warfare to compare with this.

John Frane Turner, *Fight for the Air*

With its strategic location at the crossroads of the Mediterranean, hemmed in on all sides by enemy territory and smaller in aggregate than the Isle of Wight, a mere spec on the map, the island of Malta with its tiny satellites, Gozo and Comino, represented an unsinkable aircraft carrier, naval base, workshop, and flak ship. She was in grave danger. Around her Malta practically brushed the face of hostile forces. German and Italian air bases menaced her from Italy, Sicily, and Sardinia in the north, from Libya in the south and from Vichy Tunisia in the west. Only fifty miles to the north, so close that on a clear day pilots could see Mount Etna's head from the circuit at any of Malta's three airfields, stood the island of Sicily, crammed with enemy fighter and bomber squadrons. High on a hill at Taormina stood the headquarters of the Luftwaffe Field Marshal 'Smiling' Albert Kesselring.

The Mediterranean battle would hinge on supplies as both sides reinforced themselves across vast tracts of sea and desert for the new battlefield in North Africa, where Hitler had sent General Rommel with orders to advance on Egypt and choke off Britain's supply route through the Suez Canal. Britain's supply chain was a logistical headache. To avoid enemy submarines and bombers in the Mediterranean, everything from aircraft to personnel shipped from Britain, Canada, or America had to go first to South or West Africa and then traverse the vast continent to Egypt. Engine failure over the desert meant a slow and painful death from heat and thirst for pilots transporting aircraft to bases in Egypt.

Germany's supply challenge wasn't complicated: southwards via Italy and Sicily or the occupied Balkans and Crete. But these routes were being harassed by British submarines, bombers, torpedo aircraft, and minefields laid at night from bases in Malta. Vengeance from the enemy was predictable. Yet little had been done; there had been no serious preparations. Malta had long been deemed indefensible. Only three British fighters were stationed there: ancient Gloster Gladiator biplanes with the wryly affectionate nicknames *Faith*, *Hope*, and *Charity*. They were 250 mph biplanes made of wood and fabric, first introduced in 1934 when 601 was flying Hawker Harts. Defence of the island seemed impossible, and the British government thought the island didn't have a chance and should be given up. Churchill strongly disagreed. He insisted that Malta not only be defended but strengthened as a base for offensive action against Axis shipping and air transport.

On 11 June 1940, the day after Mussolini threw in his lot with the Germans and launched his offensive, the first sirens sounded in Malta. The Italian assault was continuous for two months, with over two hundred enemy aircraft flying from bases in Sicily. What followed was a desperate struggle by Malta's defenders for the survival of its people and of the island as a war base, against immense odds, with hopelessly inferior equipment; a fight, if necessary, to the end, in isolation, hunger, and terror. Relentless bombing caused devastation. Ships attempting to bring supplies were being sunk. There was an acute shortage of fuel, parts, ammunition, medicine, and food. To the east, eight hundred miles through 'Bomb Alley', lay the sea route to the nearest Allied land base at Alexandria, Egypt. The last convoy attempting to raise the siege by this route was sunk in February 1942. That loss marked August as the starvation date for the island's defenders. The second nearest land base was a thousand miles to the west in Gibraltar, at the end of a sea lane paved with hazards.

Four Hurricanes that landed in Malta en route for Egypt were swiftly impounded and pressed into service with the fighter force, and later four more Hurricanes arrived from the aircraft carrier HMS *Argus*. As so often with aircraft sent overseas they were worn-out and wholly outclassed by the Me 109s, whose pilots were also better trained and who, as usual, employed superior tactics. On 7 March fifteen Spitfires flew in off HMS *Eagle*, the first time that Spitfires had been flown off an aircraft carrier, but still the crying need was for more fighters, for Kesselring's *Luftflotte* 2 had been transferred from the eastern front to Rome the previous December.

Kesselring had specific orders to clear a throughway north and south for General, later Field Marshal, Erwin Rommel's supplies while putting a stop to Allied east–west traffic, by seizing control of the Maltese crossroads.

If air supremacy could be denied Kesselring so would his objective. A breakthrough by Nos. 601 and 603 Squadrons, at the vortex of a battle during 'the cruellest month' of April 1942, answered the challenge.

On the day before he was due to go on leave, Bisdee was informed by Group that he was to be relieved of his present command and given No. 56 Squadron, equipped with Typhoons, handing over to Squadron Leader George Barclay. Much as he would have welcomed a Typhoon command, Bisdee remembered his solemn lecture to the squadron and their magnificent reaction. Feeling, as did Thynne in 1939, morally committed to leading the Flying Sword into action, he rang the AOC to say so. Once again, Air Vice Marshal Saul showed understanding and after a brief discussion agreed to let Bisdee remain in charge of the pilots while Barclay took command of the ground personnel. This was a disappointment to Barclay, a twenty-two-year-old veteran who after being shot down over France and returned via the Resistance was keen to assume command of a combat unit. He would do so, but his three-month tenure as nominal CO would be brief, and after taking command of No. 238 Squadron he was shot down and killed while leading a Spitfire patrol of the El Alamein area.

The first person Bisdee met when he stepped off the train at Glasgow was Ken Pawson, now one of his flight commanders. His small bundle of belongings over his shoulder, Pawson saluted smartly and then confided with a broad grin, 'There's a Yankee carrier in the docks, and they're hoisting Spitfires aboard. Could be significant!'

The carrier was USS *Wasp*, on which 601 was joined by No. 603 (City of Edinburgh) Auxiliary squadron, under the command of Squadron Leader Douglas Hamilton, bringing the number of pilots up to twenty-four. There were forty-eight Spitfires on board, and overnight more pilots arrived to make up the numbers. Deep in the bowels of the carrier, in artificially lit hangars and amid clouds of dope fumes, swarms of overalled American crewmen were already spraying the Spitfire 5Cs dark blue. To the RAF pilots who clambered down iron ladders to watch with fascination this activity, everything finally added up. They were going to Malta.

USS *Wasp* sailed during the night, escorted by a battleship, and nosed towards Gibraltar. It was its second such sailing for *Wasp*, by special authorisation of President Roosevelt at the personal urging of Churchill. It had already delivered forty-seven Spitfires, but according to John Frane

Turner in *Fight for the Air*, 'Of the 47 machines that flew off *Wasp*, one crashed in the sea on takeoff, one force-landed back on the deck as he had jettisoned his Auxiliary tank in error, one landed in Algeria, one ran out of petrol between Pantelleria and Malta, one crashed on landing at Hal Far, and one crashed off Grand Harbour.' The afternoon and evening of that day an intense raid by waves of Ju 88s was mounted in a clear attempt to wipe out these reinforcements.

There was a shuffling of pilots on *Wasp* to make up four twelves, and they were picked on the flight deck as for football teams until this was done. Bisdee retained his flight commanders, Ken Pawson and Denis Barnham, and added John ('Crash') Curry, a senior flying officer from the Royal Canadian Air Force who was actually an American pre-war barn-storming pilot from Dallas, Texas, and had owned his own flying circus in the States; Bruce Ingram, a New Zealander and son of the Dunedin Fire Brigade chief; and 'Pancho' LeBas, born in Argentina (and after the war AOC No. 1 Group).

Squadron Leader Gracie was also on *Wasp*, returning to Malta where he had already taken part in the battle, having made out his report and been specifically chosen to lead the forty-eight Spitfires off *Wasp* when the time arrived. What his thoughts were about going back to the crucible of Malta and aware of the dismal fate of previous carrier deliveries nobody knew, but Denis Barnham wrote in his book *One Man's Window*, 'Gracie is the only man among us who has been to Malta, and he looks like it.'

For her own protection, *Wasp* retained one of her fighter squadrons of Grummans, Fighting Squadron 71, led by Lieutenant Courtney Shands, and there was considerable goodwill between the Allies. The main reason for this was that the pilots of 601 and 603, knowing *Wasp* to be a dry ship, had had the foresight in their last hour before sailing to bring aboard just sufficient whiskey to last the trip without burdening their cockpit kit on takeoff. The American Executive Officer was puzzled by the unusual harmony between his officers and the RAF. 'I can't understand', he told Gracie, 'why our boys always seem to be visiting your boys in their cabins.' He wasn't told that the Americans had money but no liquor while the British had liquor but only the permissible five dollars each, and economic parity was steadily being approached through the medium of poker.

They sailed unmolested through the Straits of Gibraltar into the Mediterranean. As the air became warmer and the sea bluer, the pilots grew uncomfortably aware that they would shortly be trying their first deck take-off, in planes fitted with heavy cannon instead of the normal machine guns, and with overload tanks for maximum range. Several lectures and briefings were held to familiarise them with carrier technique and procedure, but there were bound to be some differences between the behaviour of Spitfires and Grummans. What the fighting would be like in Malta no one could imagine, and few tried, for the pilots were engrossed in their day-to-day rehearsals and detailed preparations for take-off. Take-off assumed the significance of a landing; it was the end of something, one didn't think any further. One pilot who did asked Gracie what the enemy odds were in Malta. 'If you're lucky enough to fly,' he replied without hesitation, 'forty to fifty to one.'

At dawn on 20 April, when just abreast of Algiers, the pilots bundled their kit into cockpits as mechanics scrambled over the machines giving them their last checking over. There was the final briefing. They would take off in four twelves, Gracie leading the first and Bisdee the second, followed by the other two. To avoid the heavily tiny defended enemy island of Pantelleria, which stood slap in the middle of a straight line to Malta and bristled with flak and Me 109s, a wide dog leg to the south across Tunisia stretched flying time to a little under four hours, and even this would come close to the enemy island of Lampedusa further south. Every drop of fuel was precious. Even though the Spitfires, brand-new Mark 5Cs, were fitted with auxiliary fuel tanks to extend their range, they would have to fly at low revolutions and avoid combat. In any case ammunition hadn't been loaded to allow for the auxiliary tanks. Twelve Hurricanes that took off from HMS *Argus* on 17 November had used excessive rpm and only four had reached Malta.

Bisdee chose to fly a direct route close to Pantelleria, believing fuel to be a greater cause for concern than enemy fighters and reasoning that with enough altitude they could reach the island before the Me 109s reached their height. If he was wrong about that the Spitfires should have enough fuel because of the shorter distance to be able to race at full throttle for a while until the Me 109s packed it in after a short chase over water.

Their watches synchronised and last-minute details agreed, the pilots pulled on their helmets and strapped themselves in. They were nervous. They had never taken off from a carrier before and were unaccustomed to long flights over the sea. Their navigation had better be accurate. The Spitfires were towed into position for takeoff and chocks wedged under their wheels. The pilots, after a few nervous exchanges but with a heavy sense of excitement, strapped themselves in and went again through their new take-off checklist; they couldn't help wondering how this would go. One by one the signal was given, engines exploded into life and the pilot found himself taking his first serious look at the day with the early sun framed in his windscreen, a control stick jerking in his hands as forty knots of wind rushed across the bows and past his ears.

The American crewmen impressed the pilots with their tight coordination and teamwork. When given the signal they bent forward and pulled away the chocks, their overalls plastered to their backs by the wind, and the man at the starboard wing tip gave the 'turn up' signal. To a Grumman pilot this meant opening the throttle fully against the brakes, but as this would have pitched a Spitfire on to its nose each pilot opened his throttle smoothly until the tail rose and the brakes began to slip, then released the brakes and opened up fully.

Gracie gathered his formation but set his compass 'red on blue' and initially headed towards Algeria. Bisdee set his direct course and was lucky not to be intercepted near Pantelleria, arriving with all twelve aircraft at Luqa well before any of the others after only three hours. Anti-aircraft puffs lingered over the island as they joined the circuit. 'You've just missed the nine o'clock raid,' apologised a fitter as he prised open the petrol cap on Bisdee's aircraft with a bayonet.

Of his earlier arrival Turner recalled,

> On landing I immediately removed my kit, and the machine was rearmed and refuelled. I landed during a raid and four Me 109s tried to shoot me up. Soon after landing the airfield was bombed but without much damage being done. I was scrambled in a section of four soon after the raid, but we failed to intercept the next one, though we chased several Me 109s down on the deck. Ate lunch in the aircraft, as I was at the ready until dusk. After lunch we were heavily bombed again by eight Ju 88s. Scrambled in the same section again after tea – no luck again. One Spit was shot down coming in to land and another one at the edge of the airfield. Score for the day, seven confirmed, seven probables and fourteen damaged for the loss of three Spits. The tempo of life here is just indescribable. The morale of all is magnificent – pilots, ground crew and army, but it is certainly tough. The bombing is continuous on and off all day. One lives here only to destroy the Hun and hold him at bay; everything else, living conditions, food and all the ordinary standards of life have gone by the board. It all makes the Battle of Britain and the fighter sweeps seem like child's play in comparison, but it is certainly history in the making, and nowhere is there aerial warfare to compare with this.

The tension was obvious even in the air. As the Spitfires were fifteen minutes from the island the voice of Wing Commander Woodall, the highly trusted controller of Battle of Britain fame, called out, 'Steer 081 and get a move on. Do not answer. Repeat, do not answer.' As they approached the circuit six Spitfires were sent up to cover them. Flight Lieutenant Buerling, a subsequent Canadian war hero, was stunned by his reception after landing. Before he could get out airmen grabbed his aircraft and swung it round ready for the next sortie, while others pulled him from the cockpit and another pilot waited 'drumming his fingers' with impatience. Ground crewmen pumped fuel, loaded belts of cannon shells and bullets, and fixed the radio, removing and replacing panels at feverish speed. People were shouting 'Get a move on, get a move on!' and 'Hurry up, for Christ's sake!' The new pilot jumped in and was still fastening his harness as he gunned the engine and the Spitfire started to roll forward.

A pilot came by, exhausted and eyes bloodshot, according to Pierre Clostermann DSO DFC in his book, *Screwball Buerling*, and called across to another who was waiting: 'No point in waiting for your Spit. Norman Lee was flying it and he got the chop. Get weaving, or you'll miss the mess bus, and it's a five-mile walk.'

The bastion that the fighters in their small number had to defend was an easily visible target and had strictly limited airfield facilities. Forced landings were ruled out by the terraced structure of the rocky land and there was virtually no air-sea rescue service for those who entered the water. When 601 landed they found Luqa to be even worse than they had imagined, a dusty waste without even a dispersal hut for the pilot's equipment. Deterioration due to heat cost the life of a squadron pilot whose parachute streamed futilely when he baled out.

Many of the island buildings were piles of rubble. Roads were blocked and the airfields holed. Every bomb that landed cast up a cloud of earth and dust, which lingered like gold in the sunlight until the raid had passed. Nothing could be organised, everything was improvised.

Accommodation for officers and men consisted of shelters, quarries, dugouts and the macabre catacombs. No. 601's pilots were billeted in the Naxxar Palace near Takali, where hygiene was poor and Malta Dog, a mild but incapacitating form of dysentery, was rife. Sir Hugh Pughe Lloyd, the AOC in Malta, subsequently wrote in *Briefed to Attack*:

> Starvation looked us in the face … The poor quality of the food had not been noticed at first, but suddenly it began to take effect. In March

> it had been clear enough, but in April most belts had to be taken in another two holes, and in May by yet another hole … Officers and men slept as tight as sardines in a tin; 200 slept in a disused tunnel. None had any comfort or warmth.

An item in Luqa's daily Routine Orders forbade the taking out of Malta of any official documents referring to the island's bread rationing. Bread being the staple Maltese diet, the matters of morale and security assumed prime importance under the constant threat of invasion. Only the day after 601's arrival a photographic Spitfire had returned from Sicily with alarming pictures of a levelled strip, 1,500 yards long and 400 yards wide nearing completion at Gerbini in Sicily.

At the dispersals the pilots learned that the aircraft they flew in were no longer 'theirs', nor even their squadron's, and there were no insignia or squadron letters. Each fighter was a drop in the island's pool of total resources, and everything was done cooperatively among the three airfields. No greater discrimination was being used in allocating pilots to aircraft than the dictates of serviceability called for, and this sharing was a clear necessity when one squadron might lose all its aircraft on the ground during a raid while another was airborne.

At squadron level there were no tools, no spares, no administration. All except one refuelling bowser at Luqa had been blitzed, and a carefully guarded steam roller lurked in its special protective pen to emerge furtively amid cheers at the end of each raid and try to smooth out some of the wrinkles on the runways. Morale was good, but the adverse effects of waging an all-out defensive battle were inevitable. Many ground crew were there by accident, having belonged to bomber squadrons that had had their aircraft liquidated in transit and who had been retained to service aircraft they had never seen before, in a place they had never been before, under conditions they never wished to see again. Some were 'bomb happy', and could envisage no end to the explosions around them.

The arrival of 601 and 603 had a heartening effect on morale. The pilots already there were amazed that all the Spitfires had been shed by one carrier, despite prior rumours that this would happen. Every man took a new lease of energy and fortitude from this demonstration that somebody thought their plight important after all.

If the defenders were stirred to action by this assistance, so also were the besiegers. On the day they arrived, two of the forty-one Spitfires were destroyed and fifteen put out of action by bombing. On the following morning, owing to difficulties in servicing the unfamiliar Hispano cannon, only twenty-seven of them stood ready for combat, and by the next day only seventeen.

The morning after landing, Bisdee led his section of four in a southward climb away from the approaching raiders and turned to attack them from slightly above and out of the sun. Tipped off by the island veterans he had left his hood open both to reduce the heat and to prevent the multiple scratches and specks on the Perspex, which there had been no time to remove, from being mistaken for distant aircraft. The Spitfires picked their targets from a formation of Ju 88s, some of which had already dropped their bombs, and just as deliberately the escorting Me 109s picked theirs. Bisdee took careful aim at a Ju 88's starboard motor and saw dense black smoke pour from it, its propeller slowing to windmilling speed. As he lined up the other motor his eye roved to the rear-view mirror, where he saw two Me 109s trailing in tight formation, about 800 yards away but not yet firing. Estimating that he had time for a three-second burst, he huddled behind the rear armour plating and squeezed the gun button. He didn't see what happened to his shells. There was a rapid succession of bangs as his aircraft seemed to turn inside out, the control stick stirring uselessly in his hand. Although only at fifteen hundred feet by this time he had no option but to jump, and hastily unfastening the seat harness he heaved himself bodily into the hot rush of air. Sea, earth and sky spun round him as he pulled the release handle. The chute opened with an explosive crack and tore the harness webbing from his shoulders, and only by quickly crooking his right leg round one of the loops did he avoid falling out of the harness. Two seconds later he entered the sea head first. Had the hood not been open those two seconds wouldn't have existed.

Bisdee freed his parachute, inflated his rubber dinghy and clambered in. It was now ten o'clock and the battle overhead was still new; the aircraft in his section were just making their second attacks. There was no rescue launch in sight and, having been warned that his chances of being fished out were slender, Bisdee began paddling towards the island two or three miles north. Malta's vertical, hundred-foot cliffs were slashed in sun, but gradually the triangles of shadow crept up the indentations and snuffed out the last patch of light as the afternoon wore on. At six-thirty his dinghy nudged the shaded rocks on the lee shore. Utterly exhausted, he was unable to transfer himself to the rocks. There was a heavy sea running and if his rubber raft were punctured he would drown. It was the spirited action of a naval air fitter from the airfield of Hal Far named Monck, who scrambled down a steep incline into the rough waters strewn with flower pot mines, which saved him.

As Monck dragged the half conscious pilot clear a Ju 88 dropped a stick of bombs, which covered them with earth and threw them to the ground. 'You've had a busy day,' Monck ventured as he picked himself up. 'Yes,' croaked Bisdee, 'and it's my first!' After a night in an underground bomb dump Bisdee, hardly able to walk, was driven to St Paul's Bay convalescent camp where the best part of his treatment was the news which awaited him. The destruction of his Ju 88, last seen half an inch from his reflector sight, had been confirmed by anti-aircraft shore batteries which saw it crash in the sea.

Every morning the pilots on duty were driven from their billets to the dispersal in dilapidated buses. They flew alternate days, the squadron commanders two on and one off to alternate between their two flights. Squadron entities in terms of pilots therefore remained fairly intact. The raids were as certain as the dawn. Strangely, the regularity of the bombing made things easier to handle. It was usually heavy, three times a day, allowing meals from the kitchens to be prepared and distributed in the lulls. Ample warning of the raids would be given by the long time it took the enemy bombers to rendezvous en masse over Sicily and the blur on the radar screens would be traced closely until it was time to scramble the fighters. Standing patrols would have been too costly in fuel and cut down the fighters' endurance. German R/T security was helpfully poor; formation leaders of *Fliegerkorps* 2 could be heard issuing instructions and section leaders calling to each other for a considerable time.

Group Captain 'Woody' Woodhall added to his unrivalled reputation as a controller with his deft management of the fighter defence. His skill was supplemented by the new art of 'spoofing', to perform which RAF pilots or NCOs on the ground would sit in underground cubicles with transmitters, simulating a pack of Spitfires by assigning each other targets on the Germans' wavelength. This exercise in the war of nerves, affectionately known as Pilot Officer Humguffery, was considered worth at least one twenty-millimetre cannon.

There could be no doubt that having neutralised the Royal Navy the enemy's objective was now the RAF. Almost all the bombing was now concentrated on the airfields and dispersals, and it was by no means uncommon for as many as two hundred sorties to be flown in attacks on these in the space of twenty-four hours. Before the dust could settle after an attack, squads would be feverishly at work putting things straight in order of priority and the telephone exchange would be jammed with calls for assistance. Several times the telephone exchange itself received a direct hit. On other occasions the reservoir was hit and the camp flooded with water, the latrines and washhouses demolished and the camp warning system destroyed. Aircraft in transit were liquidated and their crews transferred to the hard-pressed fighter maintenance team.

Fighters were dispersed not around the airfield but in the 'Safi' Strip. Constructed by the Malta police force and various regiments of the British army, with meagre resources and no advanced machinery, this comprised a system of winding gravel tracks through rock, scrub, terraced fields, small stone buildings, and carob trees connecting the three airfields of Luqu, Takali, and Hal Far, which allowed aircraft to take off and land at any one of them. Strung along this corridor and sprouting from it at angles where the terrain made it possible, an intricate pattern of pens housed the Spitfires and their crews. Each stretch of lane with its battery of pens was given a name that evidenced the unit instrumental in its construction, such as Dorset Road, Buffs Road, Chester Street, King's Own Road, Gunner Lane and Queen's Own Road (where 601 was dispersed).

Each pen was marked with and surrounded by petrol cans filled with earth; each pilot had his pen and crew. The fitters, riggers, armourers and wireless mechanics who slaved on the Spitfires in these pens were made up of an assortment from all three services. Army privates punctured holes with bayonets in the tops of petrol cans and refuelled the aircraft by hand as they taxied in. Each pilot was armed with a revolver, which he had full authority to use on any crew member who downed tools and fled for cover when the bombardment got too hot for him. Needless to say, such extreme action never had to be taken, the bravery and good humour of the men being beyond praise, and in course of time they even came to take charge of the pilots' firearms. The high morale generated among the pilots was in sharp contrast to the non-operational period at Duxford and Acaster Malbis.

It was essential to work at top speed in even the severest raid, for otherwise the enemy's objective might well have been achieved. Gracie, worshipped by all the men, would stroll around with well-feigned unconcern at the height of a bombing like a pagan folk leader giving advice and encouragement to the toiling maintenance, salvage, repair and bomb dispersal teams. Everything was done by teams, everything was done for The Island, every man was part of one whole, cohesive fighting unit. The common wartime phrase 'doing one's bit' might have sounded sometimes like a platitude in a large island like Britain where the individual effort was lost at the margin, but Malta was like a ship in a tempest with only a quarter of the crew she needed. When a bomb exploded frighteningly close to him one day Gracie did instinctively what any man would do – he ducked. The next day he held a short parade during a quiet spell to apologise formally to everyone for having done so. Thereafter the crewmen only ducked when there were no Spitfires to push around.

Luqa was shared by Nos. 601, 126, and 185 Squadrons, while No. 603 was at Takali together with a handful of photographic Spitfires and some Wellington bombers. Despite the occasional mixed formation a keen rivalry developed between the three airfields for the highest score, but in Easter, 1942 they put aside their rivalries to express their appreciation to *Luftflotte* 2 for the singular attention it had paid them, with 'Kesselring's Easter Hymn':

Tis Holy Thursday, let us snooker
All the bloody Spits at Luqa;
Forward Messerschmitt and Stuka –
Halleluja!

Good Friday, it is Hal Far's turn;
Blitz their crews, their aircraft burn,
Will those Tommies never learn?
Halleluja!

Easter Saturday – oh that's fine!
Make Takali toe the line,
Here a bomb and there a mine –
Halleluja!

Kampfgeschwaders rise and shine,
Mix Takali's medicine;
'Satan' works like Number Nine,
Halleluja!

The Lord is risen, ply they whip,
Smite the island thigh and hip;
Tear it off a Safi strip –
Halleluja!

'Satan' was a German high-explosive bomb, 'Number Nine' a renowned medical tablet.

Petrol, like all supplies, was difficult to bring to the island and conservation was pushed to the extreme. Air tests were impossible. The only time a repair, replaced engine, or change in trim could be tested was on an actual scramble, and an undercarriage leg might fail to retract on take-off, a cannon fail to fire when triggered, or the aircraft's trim be so bad that it required heavy effort to fly it straight and level.

After recovering from his ordeal in the sea, Bisdee led a section right into a formation of five Italian bombers. When at a good range and deflection angle his port cannon jammed, the aileron locked hard over, and his propeller began to race due to failure of the constant speed mechanism. Leaving the throttle fully open and with both hands on the stick to keep the wings level Bisdee dived, closely pursued by a grey-mottled Caproni Reggiani 2000. It was frustrating to have had no chance to shoot from an attack in which Sergeant Innes destroyed two Reggiani fighters and Sergeant McConnell one, and galling to have been chased from the scene by this gallant Italian.

The Germans were making a determined effort to clear up the job their allies had shown themselves unable to bring to a conclusion. This determination expressed itself in the ceaseless marauding of Me 109s, for which it was necessary to keep the sharpest lookout from the moment of take-off to landing, and frequently they would shoot up the island fighters when they were out of ammunition and almost of fuel, right down to the ground. A Spitfire with wheels and flaps down at a hundred miles an hour stood no chance if it caught the attention of a determined Me 109 pilot willing to take the risk. Several times an exhausted Spitfire pilot would be put through the hoop by Me 109s that had managed to isolate him and knew how to take advantage of their odds, only to be blown to pieces in his aircraft when forced to make a break for the airfield.

As in the Battle of Britain, whenever possible the Spitfires sought to avoid combat with enemy fighters and concentrated on the bombers. No. 601 made a start with its four-man snake formation but had to abandon it when the other squadrons complained it looked so menacing that they wasted time mistaking them for Me 109s.

Denis Barnham and his number two circled a pilot officer who had baled out of his machine. Fighting off four Messerschmitts that tried to sink the rescue boat they saw him slither into his rubber dinghy near the small rock, Filfola, just a few miles to the south of Malta. They saw him pulling off his boots and waving before they returned to refuel. When he was picked up by the rescue launch the pilot was found to be headless, his body mutilated by cannon shells. If the RAF felt disgust, the Maltese were beside themselves with an anger, which they made no attempt to conceal. They stood in wait with fowling pieces shooting at pilots swinging in their parachutes, and the

British pilots found it advisable to sew conspicuous Union Flags on to their life jackets.

The Italians were better behaved than their allies but incurred the grave displeasure of the Maltese, many of whom were of Italian blood. Sometimes they were unexpectedly and pathetically gallant, lumbering in tight vee formations of five or seven on their bombing runs across the island, flying impeccably but, as with pre-war RAF practice, with their eyes riveted to the adjacent wingtip so that they couldn't hope to anticipate a fighter attack. One such precise formation passed slowly over Grand Harbour and was attacked by a 601 section. Bisdee fired at one of the tri-motored, all wooden CANT 1007 bombers; the bomber burst instantly into flames along one wing, which tore away, and within seconds its remains were strewn over the grounds of the Naval Hospital. As a souvenir of this victory Bisdee took away from the wreckage a piece of the victim's tail, a rough square of green-painted canvas bearing the Royal Arms of Savoy mounted on a white cross. When he called at the hospital to make initial enquires about the wreckage a medical officer picked up the telephone to get clearance and amused Bisdee with his choice of words: 'Hello, there's an RAF type here who's called about that Bryant and May's job …'

Many Spitfires were lost in combat, their pilots' bodies never found. Malta resembled a small battlefield, impossible to crash-land on because of its criss-crossed low stone walls, and it was sheer luck if one baled out over land. A single rescue launch served the entire island, which was forty miles in circumference excluding Gozo and Comino, but as it had been so badly shot up by enemy fighters it couldn't venture far, or for long, and it tried to avoid the raids. After searching unsuccessfully for a Spitfire pilot whose parachute was last seen drifting over the sea, the log of a launch for 18 May 1942 reads:

> After we had steamed out about three miles one of the escorting Hurricanes was shot down a couple of miles ahead of us. It was while we were investigating this wreckage that Jerry got closest to us, but even then the bullets only churned up the water over a hundred feet away. As there was no survivor from this crash and still no sign of the original pilot for whom we had been called out, I decided to make for base, but on our way back we saw another fighter crash about six miles over to the westward and a parachute drifting down. We picked up this pilot within a few minutes of hitting the water and he turned out to be a Hun – a cheery soul, who advised us to get back ashore before we were hurt.

The captain of the launch had already decided to return, but as the aerial activity subsided he turned about and steered a reciprocal course for a few minutes. This proved to be a fortunate decision, for: 'We actually found the Spitfire pilot in his dinghy about nine miles from land, and the German pilot insisted on shaking hands with him as he welcomed him aboard.'

Both pilots were lucky. Ken Pawson was last seen streaking out to sea with four Me 109s on his tail. Another 'failed to return', it was never known whether he added to his score before his tanks ran dry and the warm water of the Mediterranean claimed him.

Down in Queen's Own Road there was no squadron administration; no orderly room; no adjutant, medical officer, no staff. An airman clerk wrote any necessary letters and orders in longhand with a pencil on the reverse side of disused forms. An army lieutenant combined the duties of intelligence officer and general hand, communicating with 'The Ditch' (Operations Room) by hand-cranked telephone. Every squadron commander dreamed of being freed from his desk, but in Malta the dream became nightmare reality. After seeing one of his pilots killed because of a faulty parachute, Bisdee wrote the formal next of kin notification on the back of a flimsy Unit Routine Order of a texture not unlike lavatory paper, using the taut fabric of a Spitfire's wing as a desk.

There was no opportunity to remedy affairs as raid followed raid and the island remained on the defensive. Happily there was only slight night activity, mainly nuisance raids about which everybody was too tired to care. The number of dusk raids, however, justified constant alert by the aircraft from daybreak to dark. Pilots who were off duty would occasionally supplement their half rations with 'Big Eats' purchased at local taverns or in the black market, usually scraps of ragged pork that would have disgraced the worst English wartime mess, but which were considered prize delicacies. Recreation was at first non-existent, owing to the intensity of action and the prevalence of Malta Dog, time off being spent asleep or on the lavatory. When at the end of April, however, 601 moved from the Naxxar mess to quarters in Sliema, there was a considerable improvement in health and some officers and men took to bathing in the sea. There were few parties and the acute shortage of petrol made transport scarce, while any pilot who roamed too far and missed the morning readiness bus had to make his own way on foot to his pen.

The threat of invasion intensified. By the end of April the launch airfield at Gerbini, Sicily, boasted three level runways for glider towing, and by

10 May underground cables had been laid. Almost as soon as the Spitfires landed the AOC was requesting further aid in terms of the utmost urgency.

Then suddenly the besiegers made a big miscalculation. Hitler and Mussolini, counting Malta on the verge of collapse, decided during a meeting at Berghof to put last things first. They would devote the months of May and June to Operation *Theseus*, the capture of Cyrenaica, then return to seize Malta between mid July and mid August. After this the way would be clear to invade Egypt with the forces of General Rommel, 'the Desert Fox', and take the Suez Canal.

The short, relative lull, when most of the German bombers departed for Russia, Cyrenaica and France, gave Malta a period of grace she didn't waste. Such spares and ammunition as could be flown in were not immediately absorbed in the fighting and the island was able to catch its breath. Then suddenly there was some wonderful news.

The AOC, Hugh Pughe Lloyd, summoned all his squadron commanders to a secret meeting in Mdina, on a hill overlooking the RAF airfield of Takali. He thanked them for adhering to the spirit of their pledge that they would fight if necessary to the last aircraft, and then, in an electric atmosphere, announced that they must forget this. There would be no last aircraft. Malta was going to win. Sixty-four Spitfire 5s were due to arrive from 1 p.m. onwards the following Saturday, 9 May, off the decks of HMS *Eagle* and USS *Wasp*, now steaming from Gibraltar. Absolute top priority was being given to this operation, and on The Day the entire island would be geared up for their reception. All restrictions on the use of ammunition by the anti-aircraft batteries, normally rigorously enforced with a ration of fifteen rounds per gun per day, would be lifted. The Spitfires would on arrival circle their bases at low level, while the guns would cover all three airfields and their approach lanes with impenetrable curtains of flak against any enemy who tried to interfere.

Sixty-four Spifires! The enthusiasm that this news evoked was immense, and each squadron commander raced back to inform his pilots and men. Gracie said, 'We told our pilots, ground crews and administration that we were giving them an organisation which would win the Battle of Malta, which we were at present in danger of losing. This was to be a big effort and the army would be enlisted. It would be lost if we didn't give a hundred per cent effort. The response was tremendous.'

The tremendous response was no illusion, nor was it an ordinary pulling together after a pep talk. The switch from a defensive back to an offensive rôle where victory supplanted survival as a goal gave the islanders a moral boost. Every man who was in Malta at the time testifies to the perceptible, effervescent confidence that exploded and spread across the island after the news had broken. If in the last analysis the Battle of Malta depended on material success, to those who fought it, it was a state of mind, and the battle was won with the AOC's meeting at Mdina.

Tension mounted during the long morning of 9 May. Gunners loaded their guns and waited, steel-helmeted, in the hot sun; pilots not on duty and airmen chosen for the job proceeded under instructions from the respective control towers to the downwind end of each airfield; the metal radar scanners swished steadily round in their full circles straining to pick up the first echoes from the west. A few minutes after one o'clock the taut silence was penetrated by a distant whine, which approached and became a crescendo roar as one by one the Spitfires skimmed over the cliffs from the sea, a few feet above their fleeting shadows, to bank steeply over the airfields and touchdown in stream. Some prying Me 109s were beaten off by Hurricanes and the biggest gun barrage they had yet seen. At the end of its landing run one of the waiting men jumped on to the wing of each arrival and guided the pilot to his allocated pen in the Safi strip where ammunition, fuel, oil, and food were ready. In most pens there was also a Malta-wise pilot who had the pleasure of informing the new arrival that this was the last he would see of his aircraft for some time.

Of the sixty-four machines expected, fifty-nine made it. These were refuelled, armed, checked over and completely 'turned round' within six minutes, an extraordinary achievement, and many were in action that afternoon.

A major factor in Malta's eventual turnaround was the appointment, strangely belated, of probably the world's best fighter commander, Keith Park, the widely admired commander of No. 11 Group during the Battle of Britain. Dowding had been remote and not for nothing nicknamed 'Stuffy', but Park, a veteran flyer from the First World War, made his reputation by leading from the front, flying to his Group's airfields in his personal Hurricane wearing his trademark white helmet.

The régime which began with the squealing of fifty-nine pairs of Spitfire tyres on the runways of Luqa, Takali and Hal Far, was called The New Organisation. The spirit of The New Organisation was aggression, one of forward interception. Henceforth any defensive action was to be avoided; the enemy was to be put on notice that he couldn't win. Standby was cut to two minutes, readiness to three minutes per aircraft and five for a whole squadron. Park switched to making interceptions out at sea before the raiders could reach their targets – the reverse of his Battle of Britain tactic. He also had studied the performance curves and noted that while the Me 109 outperformed the Spitfire in climb and speed though not in turning circle, there was an altitude range within which the Spitfire had a slight edge in all three respects. So he kept his Spitfires down at fourteen thousand feet, which allowed higher-altitude Me 109s to dive on them but drew them to an altitude favourable to Spitfire performance. Although it took a toll on the defending fighters it was worse for the attackers, and the improvement in kill ratio was immediate.

The grime-encrusted faces in Queen's Own Road were now cheerful and confident. Both in the air and on the ground there was exceptional camaraderie among all ranks, and although it was impossible to stray more than a few yards from the pens in case the scramble rocket should rise from the control tower, life at least became happy in the sense that only the besieged, seeing for the first time a chance to push back the besiegers, can know.

All day the pilots and ground crew lived in their pens with meals brought out, developing their own cellular life. There was 'Pancho' LeBas radiating good will and his own brand of dry humour, welcoming everything, even the Satans, as a huge joke. Denis Barnham, 'B' Flight commander and squadron artist who had painted the Flying Sword beside 601's scoreboard on the wall of the Naxxar mess, where it stands to this day, leaned against his aircraft for hours at a time with sketchbook and fountain pen; gentle by nature and hating war, Barnham fought his instincts with rare courage and, to his colleagues, appeared only as a very aggressive flight commander. 'Crash' Curry, slightly built, suntanned and moustached, was having the time of his life, although he complained incessantly that it was time the squadron got some pilots who wouldn't desert him every time he led them into a dogfight; the flying boot was on the other foot, for although a brilliant aerobatic pilot Curry was too much the individualist to be the sort of leader others could follow. To the ground crews also the brighter side came into clearer focus. With the departure of enemy aircraft they would, if time permitted, rush with their treasured slices of bread to make toast over the glowing embers of incendiary bombs scattered around the pens.

Life was still threatened but the odds were much better than before. When the island had occasion to launch its formations in any number the airmen would be heard to remark, 'The fog's rising again,' as the elliptical wings of Spitfires darkened the sky. On one day as many as sixty enemy aircraft were reckoned to be shot down.

Though the fortunes of battle were shifting in favour of the 'prisoners of Malta', as they called themselves, the problems of supplies remained as desperate as they were to Rommel in North Africa. To provision the beleaguered 300,000, whose material poverty was symptomised by the absence of razor blades, toothbrushes, soap, beer, lavatory paper, and even proper uniforms, the Chiefs of Staff elected to run the Mediterranean gauntlet again. Two convoys sailed simultaneously from opposite ends, Alexandria and Gibraltar, to divide the enemy's efforts, planning to arrive within twenty-four hours of each other. The Alexandria convoy, Operation *Vigorous*, left port on 11 June, was fiercely attacked by enemy submarines, E boats and aircraft, and having been forced repeatedly to turn about, its remnants were compelled to return short of ammunition to Alexandria. Operation *Harpoon* was a little more fortunate. Six merchant vessels, escorted by a battleship, two aircraft carriers and several destroyers, sailed from Gibraltar and braved air attacks from Sardinia and Sicily before the main escort turned back. At this point an Italian fleet made full speed towards the enfeebled convoy and after intercepting it was attacked by Malta-based torpedo aircraft. Just as the convoy came within range of German aircraft from Sicily, Malta-based Spitfires took over its protection, 601 mounting the first patrol.

Bisdee arrived over the convoy with eight aircraft early on the morning of 11 June. It was a depressing sight. On his return he reported: 'Lots of Italian warships, some on fire, others firing on convoy and us. Nineteen ships in convoy weaving like hell and taking a caning. Two cruisers very close to convoy – I flew close but discovered they were Italian. Wished I had a bomb to drop down the funnel.' Denis Barnham took the second patrol, after the third bomber and three 601 Spitfires had been shot down. The convoy was now a horrifying sight, a mess of oil patches and sinking vessels. Bisdee wrote in his log book after the second patrol: 'Convoy now down to two merchant vessels; all tankers lost, escort intact. Twenty miles south Pantelleria.'

Two of the 601 pilots whose planes were shot down baled out and were picked up by the convoy. Sergeant MacConnell failed to return.

Early next morning the two surviving merchantmen out of the entire double operation, one with its number two hold full of water, rounded Ricasoli Point and docked in Grand Harbour. As servicemen and civilians stood silently watching the historic moment, Spitfires overhead dipped their wings respectfully to the valiant sailors who had completed their mission and to their comrades who hadn't. According to the official record, the enormous cost of bringing through these two ships was one cruiser, five destroyers, two minesweepers, six merchant ships, and over twenty aircraft.

The end of the Battle of Malta was approaching, though this wasn't known, for Kesselring was beginning to conclude that the defence was too strong. Meanwhile, Hitler and Mussolini decided to postpone the invasion of Malta until September after Tobruk was taken, and a lot was to happen before then. They made this decision because the opportunity arose for Rommel to take Egypt ahead of schedule, so unexpectedly rapid was his advance in Africa.

It hadn't been possible for the British to wrench aerial supremacy from the enemy, but mastery had been doggedly denied him and great losses inflicted on his air forces. Bisdee's reaction when he was almost dragged from his Spitfire on arrival and saw the mad bustle around him had been, 'Oh, the Battle of Brit all over again!' The only thing wrong about that is that it was even worse.

It had been a harrowing experience: 3,215 alerts had been called since the siege began, and around fourteen thousand tons of bombs had fallen on the 143 square miles of Malta, Gozo and Comino, concentrated on the docks, airfields and inhabitants. Though the islands were smaller in aggregate than the Isle of Wight, this was more tonnage than fell on Britain during the entire Battle of Britain and the Blitz. There had been over three thousand air raids and 30,000 buildings destroyed. On 16 April, during a month when the sirens sounded an average of once every two-and-a-half hours, the island of Malta was awarded the George Cross.

The last ships of Operation *Pedestal* arrived on 15 August 1942, with desperately needed supplies, especially fuel. The outlook for Malta now looked bright, but only because the deterioration of the Allied position in the whole of the Middle East drew the dictators' attention elsewhere. Bisdee was called to 'The Ditch' and told that his aircraft and pilots were required by Air Chief Marshal Sir Arthur Tedder in the North African desert, where the situation was becoming grave. When the 601 pilots sailed from England in *Wasp* the ground complement had been shipped the long route around Africa via Aden to Lake Mariut in Egypt. Preparations to reunite No. 601 Squadron, consisting of twelve Spitfires and pilots, two commanding officers and a ground staff 800 miles from the aircraft, began with a flight to Landing Ground No. 07 in Egypt.

On the morning of 23 June 1942, just after the fall of Tobruk to Rommel, the Legion's Spitfires, ungainly with overload tanks slung beneath their wings, rose awkwardly from the runway at Luqa and shook themselves into formation for the flight to their next assignment. Arithmetically, the last one had been satisfactory. For the loss of three pilots an additional twenty-five enemy aircraft had been destroyed.

Overall, the battle had been a costly but monumental victory. The British lost 433 fighters, sixty-four of them on the ground. German and Italian losses were slightly higher at 462, but their losses in aircrew were much higher because most of their aircraft had several crew members. The majority of downed British pilots survived. It was losses at sea that spoke to Malta's aggressive role. Britain lost two carriers, one battleship, nineteen destroyers, and thirty-eight submarines; the Axis fleet lost 2,300 merchant ships and fifty U-Boats. The Italian transport fleet was essentially wiped out.

The RAF's commitment had been as strenuous and valiant as it would ever be, and though it may be less obvious and occupy only a small part of popular memory it writes as glorious a page in the Legion's history as does the Battle of Britain. Bisdee confessed that on learning at Digby that his squadron was being sent to Malta, he couldn't understand why the decision had been made to send such relatively untrained men, only just operational, with aeroplanes they hardly knew. By the time they were due to leave he believed that no other experience could have shaped so fine a squadron as did the eight weeks of comradeship and fervour in this confined theatre of war.

Chapter 10

Twilight of the Gods

War alone brings up to its highest tension all human energy and puts the stamp of nobility upon the peoples who have the courage to face it.

Benito Mussolini

If Rommel could take the port of Tobruk to receive his supplies from across the Mediterranean he was sure he could advance quickly to the Egyptian border and then pause briefly while Operation*Hercules* finished off Malta before he pushed on to Cairo. The massive assault on Malta had begun in early 1942, with 601 arriving there off USS *Wasp* in April.

On 21 June Rommel captured Tobruk, opening the route for ship-borne supplies. By now his advance had such momentum it seemed unwise to stop, which was why he asked Hitler for permission to continue to Cairo without pausing for the liquidation of Malta, and why it was granted. Operation *Hercules*, the invasion of Malta, was postponed until September, pressure on the island was relieved, and Mussolini arrived in North Africa with a white horse to lead the victory procession through the Egyptian capital.

Tedder wanted more aircraft, not for defence, but to maintain the air superiority needed to cover an orderly retreat by the army and its intended counter attack. His thinking was tactical and meshed with the army's plans. With the pressure now off Malta, No. 601 was a natural candidate. Two days after the fall of Tobruk to Rommel the squadron's air and ground elements were re-formed on the airstrip beside Lake Mariut. There were one Hurricane and two other Spitfire squadrons, forming together with 601 No. 244 Wing.

The desert air war took place over a vast, featureless territory for which the combatants held no special sympathy. The Allied forces were not defending their homeland or the tiny fortress of Malta. They were not defending anything; they were hand maidens of the army. There were could be no epic events, no historic RAF victories. Those would be credited to the ground forces. Living conditions for the Desert Air Force (DAF) were miserable.

The squadrons of Tedder's famed DAF were highly mobile, always close to the front line, always ready to take off at a moment's notice. Throughout 1942 and 1943 there were twenty-six 601 Squadron moves. After Maryut in Egypt, following the front line 601 moved its aircraft and pilots, ground staff and equipment to a succession of landing grounds (LGs) with uninspiring names: LG 13, LG 154, LG 173, LG 85, LG 219, Helwan, LG 154 again, LG 92, LG 21, LG 13, and LG 155, all in Egypt, followed by six more temporary bases in Libya and nine in Tunisia, prior to returning to Malta to support the Allied invasion of Sicily.

Despite America's outrage over Pearl Harbor and a natural desire to strike at Japan first, Churchill and President Roosevelt agreed at a meeting in Washington on a Germany-first strategy. The Allies knew they would have to defeat Germany to win the war, which meant invading the Continent. They also knew they were not ready. Anxious to show the American public some action Roosevelt agreed with the British proposal to engage the Germans and Italians in North Africa, where Rommel was threatening Britain's vital supply route through the Suez Canal. This was also the only area where Britain and America at that time could draw German ground forces away from their eastern front and discourage Stalin from doing a deal with Hitler, something Stalin feared his two allies might be tempted to do.

The Hurricane squadron sharing the wing with 601 was No. 73 Squadron. Almost exactly a year earlier it had lost its CO, Squadron Leader Aidan Crawley, in a strike against the Afrika Korps at Gambut. Crawley left Tangmere in April 1940 on a special mission to the Balkans as nominal Air Attaché in the Balkan Intelligence Service. After a year he applied for an active posting and was given command of No. 73 Squadron, then flying Kittyhawks in the western desert. While he was leading an attack on the enemy airfield at Gambut in July 1941 a bullet struck his glycol tank and forced him down. He was taken prisoner and flown in a Ju 52 to Greece, then to Vienna, winding up in the north compound of Stalag Luft III with Roger Bushell.

Lake Mariut's airstrip was virtually indistinguishable from the others to which 601 moved in its leap-frog movements back and forth as the German advance was halted and the front line stabilised for the Battle of El Alamein. Often there was no nearby habitation or landmark to inspire a name, simply a number. A wide plateau or an expanse of sand, the strip would be ringed with tents, vehicles, gun emplacements and trash. To the pilots fresh from Malta this seemed a profusion of equipment.

A lorry trundling slowly along would throw up a lingering wake of dust, and when a squadron of fighters opened their throttles the result was a sand storm. Such conditions were naturally hazardous. On one sorry occasion the Hurricanes of No. 73 Squadron took off across wind as 601 took off into wind on the same strip. Nothing was heard above the roar of the engines but as the swirling dust settled two aircraft were seen locked together in the centre of the runway. One of the Hurricanes had collided blindly with a 601 Spitfire flown by Di Persio, killing both pilots.

Within days there was news of an Australian Legionnaire who had left 601 at Exeter. Towards the end of 1940 Clive Mayers was posted to command a Kittyhawk squadron in the western desert. It was universally known that, as realists put it, 'It is generally not advisable to bale out over troops you have just been strafing,' so when one of his pilots was brought down by a Breda gun during attacks on Italian troops, Mayers saved him from a likely lynching by landing alongside and flying him back in his Kittyhawk. This won Mayers a bar to this DFC.

Having achieved a reputation for ground support, Mayers was promoted to wing commander and given No. 239 Wing, comprising four squadrons of Kittyhawks. He led the wing with great success on ground attack missions, closely supporting the Eighth Army, and was awarded the DSO on 28 July 1942. A week before this latest award was gazetted Mayers was leading a strike when an enemy bullet hit his glycol system, and he called over the R/T that he was force-landing in the Qattara Depression. An aircraft in his formation took a map reference and reported the details on landing, adding that the wing commander wasn't far from the advancing enemy. No. 601 Squadron was at LG 154 when, on 29 July, it was asked to search for Mayers, and Bisdee took off with Bruce Ingram as his number two. As the map reference indicated a position well behind enemy lines, the two Spitfires skirted the Qattara Depression at near-zero feet, dodging gun emplacements and fighter bases, then began a square search. For some time nothing was seen but camel bones and the twisted remains of tanks, but just as Bisdee was about to call off the search they found what they were looking for. Between the remains of two burned-out Wellingtons a Kittyhawk rested on its belly, apparently little damaged. Its hood was open but there was no sign of the pilot. With no further clue and being short of fuel the Spitfires returned to LG 154 and Bisdee made out his report. On the basis of this a later patrol was able to report having seen a solitary figure walking across the desert in the direction of the enemy concentrations.

Wing Commander Clive Howard Mayers, DFC and Bar, the volunteer from Melbourne with the funny accent, who had fought with 601 in the Battle of Britain, been saved from the Channel by Hope and an MTB, and had rescued a downed comrade in the African desert in his Kittyhawk to prevent a lynching, was never seen again. Believed to have been captured, the sad but inescapable conclusion is that he was in one of the numerous Junkers 52s commonly used for ferrying Allied PoWs to Europe that were shot down by the RAF over the Mediterranean.

For the first time in the war the squadron found itself on the side with aerial supremacy, outnumbering the enemy by three to one. It was novel and exhilarating experience. A priority of the British government, air supremacy maintained the strain on Rommel's supplies of petrol and spares, not least from Malta, and it enabled the DAF to surpass the Luftwaffe at its own specialty of Stuka-like army support.

The highly mobile fighter-bomber squadrons under Tedder were trained to bomb and strafe the enemy's troops, vehicles, and airfields during the Eighth Army's cohesive retreat, only to move back from their bases at the last moment, sometimes as German armoured cars were actually rolling on to the strip. This was done without fuss and as a deliberate tactic. Bases at the rear would be prepared to receive the withdrawing squadrons and nothing of value was left to the advancing enemy. Tedder, an excellent coordinator and diplomat, is considered by some historians to have been one of Britain's most outstanding generals during the war.

The uninterrupted hostility of the fighter-bombers not only facilitated the Eighth Army's orderly retreat but actually checked Rommel. On 4 July, his movements hampered by almost constant air attack, he was forced to switch to the defensive. Two weeks later Mussolini swallowed his enormous pride and flew back to Italy. Lieutenant-General Bernard L. Montgomery was appointed to command the Eighth Army and prepared for a counter-attack along the Alamein Line. A renewed offensive by General Rommel at the end of August fizzled out against heavy air attack, and on 19 October the DAF launched its preparatory assault.

There were many opportunities for personal initiative. Pilot Officer Llewellyn's engine was hit with small arms fire while he was attacking a troop concentration, and being too low to bale out he radioed that he was putting down on a piece of level ground among the enemy guns. Immediately, Pilot Officer Brian Kelly told Llewellyn he was coming down to pick him up. Llewellyn's plane landed on its belly, and when it stopped Llewellyn unsnapped his harness and attachments and jumped out. Kelly's Spitfire flew slowly past through a barrage of fire and Llewellyn threw handfuls of sand into the air to indicate the wind direction. Banking steeply, Kelly put his aircraft down close by as a third pilot, Pilot Officer Rowlandson, dived on the enemy positions to draw their fire. A few rifles went off as Llewellyn hastily covered the distance to Kelly's Spitfire and squeezed into the seat well from which Kelly had thrown his dinghy and parachute. With Kelly sitting on Llewellyn's lap and operating the controls between them in the diminutive cockpit they flew back to LG 172. (As late as 1957 a letter was received by the Air Ministry from an Italian soldier who had witnessed the incident. Although a gunner, he had been too surprised to fire, and recalled seeing the rescuing pilot throw out of his plane 'two cushions and a yellow pillow'.)

More than once a pilot would have a close look at his enemies. It was just after Montgomery's great victory at El Alamein and the day before the Anglo-American landings on the North African coast to the enemy's rear that six of the squadron's aircraft were patrolling Mersa Matruh as the Luftwaffe put in one of its rare appearances. Nine Stukas, three Ju 88s and nine Me 109s were seen and engaged. In the dogfight that followed three Stukas, three Me 109s and two Spitfires were shot down. The Spitfires were those of Pilot Officers Donald Llewellyn and Desmond Ibbotson. Llewellyn crash-landed beside a road and was picked up by a British armoured car. Ibbotson, his glycol system shot through, had the usual few minutes to look around for a landing ground. He chose LG 08.

Just before take-off the pilots had been given the latest army reports by the intelligence officer, Carew-Gibbs, one of whose greatest anxieties was the reputation of 'Mad Carew, Duff Gen Merchant', that some of this information, compiled by the army in haste and uncertainty, was earning him. LG 08 was among those listed as British-held. Ibbotson landed as his engine seized up, but as he climbed from the plane he was disquieted to see 'funny coloured cars' approaching, and more so when he saw the peaked caps of the Afrika Korps. His captors were good humoured and strikingly different from the German soldier of repute. They took him to a well-appointed caravan at the rim of the airstrip where he was received by a thick-set, high-ranking officer with a pair of field glasses slung from his neck, an icon Ibbotson had no trouble recognising as Rommel. Although flattered by this exchange of courtesies with this enemy commander who had earned the respect of both German and Allied soldiers for his generalship and decency, Ibbotson was deterred by the prospect of a long imprisonment and remembered that the best time to escape is as soon as possible. When darkness fell he broke from the camp and after running two hundred yards threw himself into a deep tank rut. A column of German half-tracks moved past, and when they had gone he increased his distance from LG 08 until in the morning, when a group of Bedouins helped him back to Allied lines. Within a week he was making out a sullen report for 'Mad Carew'.

Born in Leeds, Flight Lieutenant Desmond Ibbotson DFC and Bar, the 'press-on' type who would never say no to any challenge, who began his flying career as a sergeant pilot in 601, struggled with the Airacobra, and became a flight commander in May 1944, died later that year in an unexplained flying accident in Italy.

The seven months from October 1942 to May 1943 were crucial. The German defences at El Alamein were penetrated on the night of 23 October, and by 12 November the Eighth Army was in Tobruk. Rommel's bid for Egypt had failed, and when the Anglo-American invasion forces landed at Casablanca, Oran, and Algiers he was pincered between forces to the east and the west. The next important victory for the Allies before driving the enemy out of Africa was at Tripoli. Meanwhile, the forward squadrons of the DAF continued to harass the Germans' retreat to Tunisia. From now on air supremacy would be the main factor in Allied success.

Hitler's surprise assault on Russia had lifted the pressure on Britain and sparked the first hope of eventual victory. The Japanese attack on Pearl Harbor, which kicked a sleeping giant and brought the United States into the war in December of that year, suddenly made it likely. Finally, the American victories at Coral and Midway, Britain's at El Alamein, and Russia's at Stalingrad, all within a period of six months, shortly followed by Germany's defeat in the biggest tank battle of all time at Kursk in Russia, made it virtually certain – provided, of course, the Allies held together. Even the German high command knew the writing was on the wall. It was now a simple matter of staying the course to the end. Simple, that is, except for those still doing the fighting.

The ground over which the Eighth Army advanced was strewn with broken and abandoned equipment and the 'graveyards' of aircraft. Frequently only elementary 'cannibalism' was needed to make some of the aircraft serviceable, such as a Messerschmitt with a flat tyre alongside another with a damaged propeller. By ingenuity and hard work No. 244 Wing availed itself of an He 111 and a Fieseler Storch and used the latter, a slow high-wing aircraft, for communication between airfields, contrary to all regulations. The Legion acquired its own FW 190, an Italian CR 42 biplane, and a Stuka. Only the grey-mottled Stuka had what appeared to be potential. It was given RAF roundels, generous yellow wing bands and a red flying sword, with semi-conforming squadron code letters 'UF-?' and flown forward with successive advances, its crew of two making the necessary arrangements for the squadron's comfort before the Spitfires arrived – again completely against regulations. The practice stopped abruptly when corrosion was detected in the Stuka's wing spar.

Canadian Ray Sherk was an early product of the EATS, soloing at Saskatchewan on Christmas Eve 1940 in the standard Tiger Moth trainer and winning his wings there on Harvards with an above-average rating. He graduated on Spitfires at the Operational Training Unit (OTU) in Heston, England. The RAF posted him to No. 74 Squadron in Egypt, sending him by the circuitous route of Durban, South Africa, to avoid the hazardous western Mediterranean. The cargo ship conveying his and other Spitfires was sunk, and he had to fly Hurricanes for a couple of months in Egypt and Palestine with No. 73 before, in August 1942, joining 601 and its Spitfires. The 601 CO was Squadron Leader Arthur Clowes, DFC, DFM, an ace with eleven victories who had served with No. 1 Squadron during the Battle of Britain.

Sherk and 'Crash' Curry took off in their Spitfires on 29 September 1942, led by Squadron Leader Peter Matthews of No. 145 Squadron. After flying westward out to sea they turned inland and attacked troops and docks at Mersa Matruh, a radar station on a hill, a troop encampment, a military headquarters, a supply dump, a train and a motorised column. Finding no further targets of opportunity they returned to the train, which had already been brought to a halt, and after hitting the locomotive left it gushing steam and the trucks in flames visible for twenty miles. As they turned to go back Matthews spotted an aircraft to their left, which they readily identified as a Ju 52. Sherk, in the echelon port position and closest to the target, fired first. There was some ineffectual return fire from the rear gunner, then the Ju 52's starboard wing and fuselage burst into flames. Its pilot tried to land as the other two Spitfires followed up the attack. The Ju 52 crash-landed and burned. Seeing nothing more to be done, Matthews gave the order to return to base. Minutes later Sherk's engine failed. He force-landed in the Qattara Depression and was taken prisoner by the Italians, who flew him to a PoW camp in northern Italy.

When Italy capitulated in September 1943 all Sherk's camp guards walked out, leaving the place unguarded. Knowing the Germans would soon take over, he and several other escapers took to the hills. There he was sheltered by partisans and passed from village to village over the next few weeks until he reached the advancing Allies. When he learned that No. 601 was near Foggia he made his way there. Not without a sense of the occasion he opened the flap to the mess tent and called out, 'Hi fellas, I'm back!' Unfamiliar faces stared at him. Someone muttered, 'And who the hell are you?'

Sherk returned to England and resumed operations with No. 401 Squadron, flying Spitfires over France. In March 1944 he again had an engine failure, baled out, and avoided capture with the help of French patriots who smuggled him into Spain. The BBC announced his safety with the coded message 'The sky is blue'.

Rommel, his bridgehead contracting, withdrew to the fortified Mareth Line, built by the French for defence against for Italians.

By the end of February 1943 some RAF bases were so close to Rommel's army that they were within range of his artillery. Hazbub Main had repeatedly complained to the army about spasmodic shelling, but despite some vague assurance that Ghurkhas would be sent at night to wipe out the emplacements nothing was done. On the evening of 1 March the bombardment began in earnest. The first shells struck the camp at ten minutes to five, puffs of smoke on the neighbouring hills indicating a large number of guns. The shelling was incessant and accurate. Without hesitation the No. 244 wing commander ordered all serviceable aircraft to El Assaa and Zuoara, and they made their take-off runs through heavy fire. A shell exploded under Sergeant Reynold's plane as it was gathering speed, destroying it and seriously injuring him. Well into the hours of darkness the unserviceable planes were manhandled out of artillery range.

A few days later the Eighth Army breached the Mareth Line and met up with the American Second Corps. No. 601 returned immediately to Hazbub Main. One by one, Cap Serrat, Bizerta and Tunis fell to the Allies, and in the first half of May 1943 the entire Axis forces in North Africa surrendered. The way was open for the invasion of Sicily and Italy. As the depleted Luftwaffe withdrew, leaving to the army commanders the ignominy of surrender, the Legion lent support to the bombardment of Pantelleria and Lampedusa, the island fortresses in the Mediterranean which had been the bugbear of carrier-born relief aircraft for Malta.

Murdoch Matheson, known throughout his service as Bill, was another product of EATS, from another distant part of the British Empire. His home was a family sheep and grain farm in Wycheproof, a tiny community that might today be condescendingly called a flyover town, in the area known as The Mallee, a dry, shrub covered, sparsely populated area of Western

Australia devoted to agriculture. His family had owned the farm, growing crops and running a few sheep, since Bill's great grandmother, the daughter of Scottish parents, bought 320 acres under a government plan to settle the land. As the farm passed from generation to generation it grew to 840 acres.

Young Bill Matheson was happy. He had no yearning to fly or go abroad, but he was to do both. In 1940, anticipating call up at the age of eighteen, he looked at his possibilities and after reading about EATS chose that path without any clear idea of what to expect. The global scale and logistical complexities of EATS were impressive. There can be no question that without the scheme this potentially outstanding fighter pilot would have been lost to the air force. After a medical exam and ground tests in Melbourne, Matheson passed as a pilot candidate, was kitted out with flying clothes, and posted to Rhodesia for training. After tens days' leave he sailed from Sydney on the *Queen Elizabeth*, converted to a troopship, on which he made his first and last encounter with bed bugs. His route was circuitous. After docking overnight in Fremantle on Australia's west coast the ship continued via Ceylon and through the Suez Canal to Alexandria. Here everyone disembarked for several weeks with nothing to do, until they were embarked again and the *Queen Elizabeth* steamed back through Suez via Aden and Mombasa to Durban, South Africa.

This was where Matheson acquired his wings on Tiger Moths, with no difficulty, in the benign climate of Rhodesia, winning a contest for first pupil to go solo after seven hours and ten minutes of dual instruction. He was one of only two cadets out of fifty selected for a commission. Then it was off to Alexandria again by another ship through the Suez Canal to the DAF.

At introduction to the OTU near Alexandria, Matheson's course was given a pep talk. They were no longer in Training Command and could fly as low as they liked; they could even beat up the landing field – provided that they did at least two climbing rolls afterwards, otherwise they would be in trouble. The fledgling pilots graduated on Harvard trainers, single-engined stepping stones to fighters, and transferred to Spitfires. Confirming the belief that the further away from Britain the older the aircraft, the Spitfires were ancient Mark 5Bs with a manually operated undercarriage. Two of the instructors were Bruce Ingram and Pancho leBas, both of whom had flown off USS *Wasp* to Malta. Now in the DAF, Ingram replaced the formations of six, amazingly still in use, with the Luftwaffe's tactically far superior finger four. In a reversal of 1940, however, new RAF pilots such as Malleson were arriving thoroughly trained while it was now the Luftwaffe's turn to rush in green replacements. At the urging of Ingram and LeBas Matheson applied to join the Legion.

While eating in the pilots' mess, consisting of two small marquee tents joined together, Matheson noticed a man sitting alone at another table, talking to himself occasionally. He was quietly told not to appear to take any notice; it was the CO, Squadron Leader John Taylor, DFC.

The squadron returned to Malta in March 1943 under Taylor to help cover the projected landings on Sicily. Matheson's junior status required him to follow as a passenger in a Bisley bomber to Tunis, followed by a sea crossing to the island in the cramped double-decked bunk of an American landing craft. The island had changed. Supplies had been brought in by convoy and there was a greatly increased capacity for Churchill's goal of waging offensive war. The Americans had built a landing strip on Gozo, there was a spacious new operations room, and the radar equipment had been replaced. Luqa was the centre of a strange activity, unhampered by enemy bombing, that of preparing as before for invasion – but 'the other way'.

On the night of 9 July 1943, the sky over Malta throbbed with the sound of Dakotas towing 137 gliders to the dropping zones in Sicily, and with Hurricanes flying out to attack the enemy searchlights. The seaborne elements were already under way, and in the early hours of next morning the British and Canadians landed on the eastern and the Americans on the western end of Sicily. Ten of 601's Spitfires took off to cover the landings and arrived over the beaches.

Hundreds of amphibious vehicles were strung along the Gulf of Noto, where the Eighth Army was landing, the shallow waters freckled with infantrymen wading ashore while further out to sea larger ships formed protective arcs.

The fighters were stepped from ten up to twelve thousand feet and there was no opposition until, over the Gulf of Augusta in Sicily, eight Stukas were seen heading for the landing beaches at eight thousand feet. Taylor led the Spitfires in, held his own fire until very close at about a hundred yards, then was hit by return fire and pulled sharply upward. He told his formation he would try to make the Pachino beaches, now in British hands, but these were thirty miles away and within minutes he reported that his engine had cut. Now out of sight, his last words were that he had found a beach where he would try to land. Advancing British troops later found the charred body of Squadron Leader John S. Taylor, DFC with fifteen victories, in the burned-out remains of his Spitfire on a beach just south of Syracuse.

(Such, at any rate, is the official report. Contemporaneous accounts by the squadron's ground crew describe Tayor's Spitfire crashing into a stone wall on the airfield. It is possible that Taylor changed his plan when he saw the airfield, then typically overshot because of his dead stick and hit the far wall. Official reports are not always accurate. Both accounts are credible, and one of them is true.)

Three days after the landings and following the practice of moving with the front line, 601 transferred to Sicily. The new CO was a Polish officer of the regular RAF, Stanislaw ('Stanley') Skalski, DFC, who had flown with various RAF Polish squadrons. He had been personally appointed by AVM Harry Broadhurst, commander of the Western DAF, who wanted to replace tired British and Commonwealth COs and said he knew Skalski could do the job. He certainly could: 601 pilots considered Skalski the best fighter leader they had ever known. He was also unusually humane. While in the Polish air force fighting the German invaders in 1939 a Henschel HS126 reconnaissance aircraft that Skalski's formation attacked was shot down. Skalski landed beside it and tended to the wounded crew members until they could be taken to hospital.

The Germans launched air raids on the newly established Allied airfields, and at half-past ten on the brightly moonlit night of 11 August they bombarded the squadron's base at Lentini West, first dropping flares that illuminated the whole camp and then high explosives and anti-personnel bombs for a period of forty-five minutes. There was heavy damage to aircraft and equipment and four men were killed, three of them pre-war Legionnaires.

For all the Germans' determination, the Italian will to resist had been broken. Mussolini was forced to resign (and ultimately to be lynched by his countrymen) and Marshal Badoglio led the Italian nation. On 3 September 1943, four years to the day after Britain's declaration of war, the armistice was signed at Reggio. Announcement of this development was withheld for five days, but on 9 September the squadron's pilots climbed from their grimy Spitfires at five-thirty in the afternoon, after covering the Eighth Army's landing at Pizzo, to learn that Italy had surrendered unconditionally to the Allies. The terms of this surrender called for the immediate end of hostilities by the Italians, the free use of all Italian air bases and ports, and transfer to the Allies of the Italian fleet and air force.

Marshal Badoglio fled to the Allied cause and declared war on Germany in Italy's name.

It appeared a very novel development to the British forces, particularly those that had fought in Malta, to welcome Italy as an ally. Next day's duty was an example of the weird things that happen in war. In shifts of four, the squadron's aircraft escorted the portions of the Italian fleet from their home ports as they sailed in humiliation for Malta, fourteen warships in one formation and a convoy of one battleship and three destroyers comprising the other.

The wheel had turned full circle for Italy. For the Flying Sword squadron it was now a far cry from the Safi Strip at Luqa to the olive groves at Lentini West, and the sword was sheathed for a while as the wings opened protectively over the first Italian ships to see, at long last but in unimagined circumstances, for Grand Harbour, Malta.

The Germans were not so easily beaten, and the Wehrmacht didn't retreat to the north as expected, despite the Luftwaffe's virtual withdrawal to the Eastern Front and a supply situation so critical that oxen were a standard mode of transportation. For the next twenty months 601 moved forward with the line, assuming temporary residence at numerous undistinguished landing strips, seeing the coming and going of four more commanding officers before the war ended.

'Ned' Grosvenor would have been intrigued to see his Legion in Italy. By now a pilot in 601 was more likely to be of foreign birth than British. Bill Matheson listed the nationalities of its thirty pilots at one point: thirteen from Britain, six South Africa, five Australia, three New Zealand, two Belgium, and one the USA. The far-sighted EATS had made a big contribution. Devised in the dark days before the Battle of Britain, it took time, but its reach was wide and deep.

In the first few weeks after the capitulation of Germany's ally the squadron, undeceived as to the efforts which lay ahead following the collapse of a nation that had been hesitant and unimpressive in combat, set itself a short term objective – destroying its two-hundredth enemy aircraft. On the day of Italy's surrender the score stood at 198. The next plane destroyed was one of the squadron's Spitfires, which hit the muzzle of a Bofors gun in a mock attack, killing the pilot.

Allied air supremacy was complete; almost nowhere was the Luftwaffe seen, its fighters having been drained away to the Russian front and for home defence against mass British and American bombing raids. No. 244 Wing became a long extension of the advancing army's artillery, hitting storage dumps and communications land leaving a trail of destruction behind it. On one of these armed reconnaissance flights a Ju 88 was shot down by Bill Matheson, leaving one more to go for the squadron's two-hundredth.

As 601 moved forward from Foggia to Triolo two Spitfires crashed on the landing run, killing one of the pilots. Ten days later four Me 109s were seen by four 601 aircraft patrol, which succeeded in slipping unobserved on to their tails. The leading Spitfire was within six hundred yards of the target and had still not been seen when it was itself attacked from the beam by a well intentioned Spitfire from another squadron. In the resulting confusion the German pilots perceived their danger and escaped.

It was a month before enemy aircraft were seen again, and that came as a surprise. Seven Spitfires were patrolling the Pescara area at midday when, without warning, six Me 109 fighter-bombers dived steeply through the middle of their formation, dropped their bombs and made off for Pescara. The Spitfires gave chase at full boost, cut off the corner as the Messerschmitts turned, and damaged one with a long burst from the formation leader. The leader's number two followed up and shot the plane down in flames.

The two-hundredth victory was that of Flight Sergeant Eid, a Belgian who had escaped from his country in the early days of the war and enlisted in the RAFVR. He never increased his score. Taken off operations at the end of a tour, Eid contented himself with practice flights while awaiting posting. He took up a newly repaired Spitfire on test, but had just retracted his wheels when for some reason he tried to land again off a tight circuit; on the final turn his aircraft stalled and spun, killing him as it struck the ground two hundred yards short of the runway.

The Allied advance continued, but slowly. As the winter of 1943 drew in with heavy winds, rainstorms, slush and persistent cold, the weather became unsuitable for all forms of warfare. Flying, driving and even walking became impossible for days on end; tents were destroyed by gales and aircraft wings had to be covered to prevent the formation of frost. By Christmas no one any longer expected a quick victory in Italy. But in the new year, during the landings at Anzio for an assault on Rome, there were days when the clouds lifted and the Luftwaffe actually put in some intensified effort, providing more action than at any time since Malta. In this month four enemy aircraft were destroyed for the loss to the squadron of four but no pilots lost.

The landings south-west of Rome eventually brought success, and at 9.15 on the evening of 4 June 1944, two days before the invasion of Europe began in Normandy, Rome fell to the Allies. Kesselring retired to the north behind the Gothic Line, protected by the western curvature of the Apennines from attack along the western coast north of Florence. While he was left to believe that this was where the next assault would occur, elaborate plans were made by the Allied commanders to transfer secretly the weight of their forces to the low lying Po Valley in the east. All moves of the DAF were thoroughly obscured. Squadron emblems and markings were removed from aircraft, and 601's were stripped of their Flying Swords for the transfer from Perugia to Loretto on the Adriatic Coast, as No. 82(F) Unit.

At Loretto, No. 601 backed up the Eighth Army with attacks on specific targets identified by VHF installed in army lorries, jeeps and trailers. The Spitfires would circle in a 'Cab Rank' until directed by ground forces on to individual targets and pockets of resistance, both stationary and moving. Only total air supremacy made this luxury available to the army. When called in to attack, the fighter-bombers strafed with cannon in their dives, released two five hundred-pound bombs from two thousand feet, then expended their remaining ammunition on any further signs of life. It was brutal, bloody work.

One perquisite that fell to the side owning the skies was a first class air-sea rescue service. A 601 pilot, shot down by flak, found himself in his dinghy, surrounded by mines, in the mouth of the River Po. Four Spitfires took off to cover his rescue, being joined by an impressive working party of one Warwick, one Mosquito and two Walrus amphibians. As the Spitfires circle and dived on the enemy guns, the other aircraft made a detailed survey of the task and formulated a plan. One Walrus alighted on the water at the edge of the minefield and the Warwick dropped a lifeboat to it; the Walrus crew jumped into the lifeboat, rowed it through the mines to the stranded pilot, returned with him to their aircraft and took off. Throughout this operation the rescued pilot, despite hostile fire, mines and the air display above his head sat happily eating his survival rations.

The security ban was lifted and the Flying Swords sprouted on every surface that would take paint. As though in celebration, No. 601 took the Wing record by flying sixty operational sorties and dropping thirteen tons of bombs in one day. Messages of appreciation poured in from army commanders who were delighted by the exactitude and thoroughness of the close support they received. The technique was to dive from ten thousand feet at an eighty-degree angle, point the nose straight at the target and hold it steady, then release the bomb and pull up at three thousand feet. Even without a bomb sight their accuracy was surprisingly high. The Spit bombers had indeed out Stuka-ed the Stukas. When Field Marshal Kesselring was removed to the Western Front and succeeded by General Von Vietinghoff, the latter's opinion of them was emphatic: 'They hindered essential movement; tanks couldn't move; their very presence over the battlefield paralysed movement.'

But the risks were high too, and so for No. 601 Squadron was the casualty rate. Group Captain Hugh ('Cocky') Dundas DSO, DFC, the wing leader, later said, 'The last four to six months of the war in Italy were the most dangerous and terrifying period of the war. Not only were the Germans now extremely accurate in their ground-to-air firing, but a consignment of 500-pound bombs that we used had faulty detonators and at least two of our pilots were blown to smithereens in bomb dives.'

On 14 April 1945 the Fifth Army attacked the centre of the Gothic Line and by a pincer movement with the Eighth Army marched with it into Bologna, trapping many thousands of German soldiers. The escape routes were cut off; there was nothing more the Germans could do but surrender. This Von Vietinghoff did on 24 April. The cease fire was set for 2 May.

In the last month of the war, No. 601 Squadron won the DAF record for the most operational hours flown by a single squadron at any time. Never had a thousand hours been topped. Anxious to leave no room for doubt, the squadron flew every possible operational mission until the last. The last one occurred two days before the cease fire, when Flight Lieutenant O'Halloran, Flying Officer Ross and Flying Officer Hallas took off at three-thirty in the afternoon to strafe motor transport in the Conegliano area. This was routine work, and they destroyed four vehicles before deciding to call it a day. Hallas's aircraft had been stuck by anti-aircraft fire but he didn't consider the damage serious. While returning, however, he told the other pilots that his oil temperature had risen to 120 degrees and that he might have to bale out. Hallas wasn't seen to leave his aircraft, which crashed on a ridge four miles east of Rovigo. He was the last No. 601 pilot to die in action.

During the final month of the war in Italy, No.601 Squadron achieved a record 1,082 operational hours in the month – the first time a fighter squadron had exceeded the thousand-hour mark in this Group. The number of pilots killed or lost by baling out or force-landing was high. Flying Spitfires with heavy bomb loads against the dug in, well trained, and disciplined German army had proved extremely dangerous.

By war's end, Bill Matheson was the longest presently-serving pilot in 601, with 194 operational flights and 300 operational hours. He had fought in North Africa, covered the Allied landings at Anzio, and escorted American bombers on the controversial raid on Monte Cassino in a failed attempt to draw German troops away from Rome, with an overhead view of the medieval monastery being pummelled to rubble. Having served two voluntary extensions of fifty hours each beyond his tour of 200, Matheson applied for another extension but was told 'enough is enough'. He was made to take a leave, to recuperate and refresh, and when he saw the Spitfires being fitted with bomb racks he wasn't sorry to do so. Air-to-air combat was more within a pilot's control and largely dependent on skill and courage, but air-to-ground was a pure gamble. Not only could the pilot not see where the lethal anti-aircraft guns were but there was always the danger of having to bale out or force-land near troops he had been strafing. Flight Lieutenant Bill Matheson took leave and then returned to his family's farm in Wycheproof.

At Treviso in Italy, in May 1945, No. 601 Squadron was disbanded. Its contribution to a long and bitter fight for survival and ultimate victory, in the only material terms that can be measured, was over 200 enemy aircraft destroyed in combat and six on the ground, as well as 229 motor vehicles and several Tiger tanks.

Chapter 11

The Fate and Fame of Roger Bushell

I had greatness thrust upon me … Nothing is ever closed to me, not ever.

Fl/Lt Bushell

Roger is always hiding his light under himself …

Mike Peacock
Entries in the 601 Squadron Line Book, 1936

In 1963, a widely acclaimed film titled *The Great Escape* told the exciting story of a tunnel escape from a prisoner-of-war compound for Allied airmen in southern Germany. Richard Attenborough played the part of RAF Squadron Leader Bartlett, who as 'Big X' master-minded, organised and oversaw a massive tunnel escape from a German PoW camp. His role was somewhat overshadowed by that of Steve McQueen as a downed American pilot in a role that established him as a world class motion picture actor. The film was based on a book by Paul Brickhill, an Australian-born Spitfire pilot who was shot down in 1943 over Tunisia and was also a prisoner of war at Stalag Luft III. The escape story is one of rare imagination, organisation, and courage deserving to be shown as a motion picture. As such it was told well. But in reality no role overshadowed or even matched that of the man who was 'Big X', and he was an RAF pilot named Squadron Leader Roger Bushell. Nor was there a single American in the real escape.

When Bushell arrived at Croydon in 1940 as a squadron leader to re-form and command No. 92 Squadron, its role was night flying on Blenheims. He was deeply sad to leave 601 and took two Legionnaire friends with him: 'Paddy' Green, also a South African and a bobsleigh champion, and Munro-Hinds. Before long the other 'Paddy', Flying Officer Byrne whom Bushell had helped defend after the Battle of Barking Creek, also transferred from 601.

Putting together a new squadron was a new challenge to which Bushell rose with new-found skills of planning and organisation, which would later serve him well in the most important episode of his life. Mike Peacock's witty comment that 'Roger is always hiding his light under himself' was more than a play on words; it observed Bushell's unconcealed ego, his self-advertised over-confidence. In sport, in his profession, and in flying, he was never far from the edge. By the time No. 92 Squadron was operational and equipped with Spitfires, fun-loving Roger Bushell was all business. Instead of tussles with authority, now he *was* authority. He was disgruntled at receiving green pilots who hadn't been sent to an OTU, and one of these was Geoffrey Wellum, a newly minted eighteen-year-old pilot officer fresh from basic training.

Wellum describes his first encounter with Bushell at Northolt in his book, *First Light*. He is interviewed by the adjutant, who asks how many hours he has flown. When told a hundred and sixty-eight with ninety-five solo the response is 'Jesus H!' The adjutant arranges for Wellum to see the CO. 'His name is Bushell, Roger Bushell. He was an Auxiliary of long standing and in private life is a barrister. He's quite a bloke to serve under. Don't be put off by his manner. To him his squadron and his pilots are everything. Once you are accepted into his squadron you will never find yourself alone.' Bushell makes a powerful impression on Wellum: 'The man at the desk has a tremendous personality. He's a pretty hefty chap and just plain ugly in a pleasant sort of way. He stares at me for some seconds. If this is an example of a typical barrister then God forbid that I should ever end up in a court of law. He has my log book open in front of him. It cuts no ice.'

On the morning of 23 May 1940, Bushell led his formation on a patrol of the Dunkirk area but saw no enemy aircraft. In his formation was one of his new flight commanders and a future ace, Flight Lieutenant Stanford Tuck. On a second patrol over Boulogne Bushell hurled himself into a circling formation of Me 110s, damaging two before being shot down himself. He crash-landed in a field that he mistakenly believed was in Allied hands. Wellum, who survived the war as a successful Spitfire pilot, believes Bushell acted recklessly in throwing himself into a hornet's nest of Me 110s.

Bushell stared dumbfounded at the pistol pointing at him and realised that if he hadn't hailed this bloody motor cyclist he wouldn't be a prisoner-of-war. He, a prisoner, on his first engagement and after only eight months of war!

He was first posted killed in action. Later, in a personal letter to his parents in South Africa, an Air Ministry Group captain wrote that he was a prisoner and though there were as yet no details he hoped 'he is well, and making such a nuisance of himself to the Hun that they will deeply regret having taken him prisoner'.

Bushell fulfilled that hope. Neither then nor in the years to follow did he go quietly. He waged a relentless, private war against the Germans; he escaped, was recaptured, escaped again and was recaptured again – and escaped yet again with insuppressible doggedness. He knew that every captured officer was entitled by the Geneva Convention to attempt to escape. But there was more to it than that. Simon Pearson records in *The Great Escaper* that early in the war a highly secret unit of British Intelligence was created and code-named MI9. Believing that PoWs could be an intelligence source, MI9 briefed selected squadron commanders on a coding system for letters sent home should they become PoWs, with which they could convey information about enemy troop dispositions, military and industrial facilities, and civilian morale – none of which, it must be realised, was compliant with the Geneva Convention. They taught escape and evasion methods and instilled into their charges the duty to escape. Nobody took this duty to heart more seriously than Bushell.

The first place to which he was taken was Dulag Luft, a transit camp for aircrew prisoners near Frankfurt. After a period of solitary confinement, Bushell made a survey of the camp. In the playing field, and just outside the compound wiring, there was goat in a kennel. If a hole were dug in the floor of the kennel and a trapdoor fitted to support the goat, a man could jump in it and remain concealed from the sentries as the prisoners returned from the playing field after exercise. The hole was dug by relays of prisoners hiding in the kennel one by one, the sand being taken away in vessels used for feeding the goat. If the guards had counted the number of times the goat was being fed their suspicions would have been aroused, but they didn't.

Bushell planned to hide in the kennel on the evening before a planned tunnel escape involving a number of prisoners. He would climb the single wire surrounding the sports field as soon as it was dark, and when his escape was discovered his pursuers would think he had escaped with the tunnellers and not with a twenty-four hours start over them. On the prospect of staying in the kennel someone asked, 'What about the smell?' and another prisoner gave the predictable answer, 'Oh the goat won't mind that.'

It was an easy matter to falsify the roll call, and Bushell got away from the camp smoothly. With his fluency in German and experience of the winter sports areas he set course for Switzerland, travelling by day in a civilian suit bought from one of the guards at Dulag Luft. He was able to engage safely in brief conversations, and navigating with the aid of guide books purchased from shops along the way he went to Tuttlingen by express train, and from there to Bonndorf by suburban line. His plan to throw the Germans at Dulag Luft off the scent was entirely successful, for none of the eighteen men who escaped by tunnel got farther than Hanover before being arrested, by which time Bushell was beyond the radius of search.

From Bonndorf, Bushell reached on foot the point he was making for, a few kilometres from the Swiss border. Things had gone almost too well and, being aware of his habitual over confidence, he sat down for two hours and made himself generate caution for the last decisive stage. He had the alternatives of waiting for nightfall, with all its problems, or of bluffing it out by daylight. He chose the latter.

In the border village of Stühlingen he was halted by a guard. Pretending to be a drunken but amiable skiing instructor, Bushell was being conducted towards a check point for an examination of his papers when he broke loose and bolted, dodging bullets, into a side street. This proved to be a cul de sac and he was run to earth within a minute. He served a punitive sentence in a Frankfurt gaol, intended to soften his morale; but on being moved to Barth, near the Baltic coast, he escaped again with a Polish officer. The two men separated, and Bushell was stumbling along a road near the concentration camp at Auschwitz on a dark night when he blundered into a sentry he hadn't seen, knocking him to the ground. With an instinctive courtesy he helped the soldier to his feet, handed him his rifle and said, 'Sorry!' The game was up once again.

It was decided to move this troublesome officer to a new camp, and he was herded into a cattle truck with several other prisoners and taken from Lubeck to Warburg. What awaited him there Bushell didn't wait to see, and with five others prised open the truck's floorboards and dropped on to the track as the train was moving. One of the prisoners dropped on to the rail and lost both legs as the truck rolled over him.

With a Czech pilot named Jaroslav Zafouk, Bushell reached Czechoslovakia where the Resistance boarded them both with a courageous family in Prague. Bushell appreciated this limited freedom and would take daily walks in civilian clothes around the city while waiting for the Resistance to complete arrangements for his transfer to Yugoslavia. But the assassination of the Nazi tyrant Reinhard Heydrich activated a ferocious response, including the massacre of a whole town at Lidice. In Prague, hostages were shot and a house-to-house search for suspects of the assassination was conducted.

Bushell didn't know it, but he was a marked man. According to Pearson he had had an affair with a girl named Blazena living in the household that sheltered him but had refused to agree to eventual marriage. Distressed by this rejection, she inadvertently betrayed him by telling a trusted friend, who was actually an informer, about the two fugitives the family was sheltering. All, including Blazena, were arrested, tortured, and shot.

Bushell was arrested and taken for interrogation to a Gestapo facility in Berlin. He wasn't tortured but treated harshly and heard the sounds of others who were. His cell was one of a number on either side of a corridor, and when they had locked his door and withdrawn he put his face to the grille and asked softly: 'Is anybody here British?' A voice four cells away in the direction of the latrines replied, 'Yes, Flight Lieutenant Marshall, RAF.' Marshall, who had known Bushell before the war, was also an escaper and had just been re-captured. Conversation between the two was restricted to furtive whisperings of a few seconds' duration whenever Bushell passed Marshall's cell. It took several days for Bushell to explain that he was refusing to admit his identity for fear of repercussions on the Prague family, which would be telling the same story as his. He was tormented by the thought of what would happen to them. One evening he whispered that he had left a note in the lavatory. When Marshall found it tucked behind the cistern it contained Bushell's service number, rank and full name. 'They are going to shoot me,' it stated; 'Please pass full particulars to the Red Cross.'

Bushell learned that the Prague family had been executed and admitted his identity. He could at this point have been lawfully executed for wearing civilian clothes and associating with members of the Resistance, but the Luftwaffe took him under their protective wings and sent him to Stalag Luft III. It was here that he received his ultimatum from the friendly commandant, Baron Oberst Von Lindeiner: 'Be warned; if you escape again you will be shot, and there will be nothing we can do to save you again.' But by now, with memories of the treatment he heard others going through in adjoining Gestapo cells and the knowledge of what had happened to the brave Prague family, Bushell's determination to fight on was unshakable.

The Luftwaffe treated its prisoners decently, even generously. Hundreds of thousands of captured servicemen in Germany adjusted themselves successfully to prison-camp life, studying philosophy, learning languages, throwing themselves into camp sports and theatricals, and by so doing found life tolerable and in many cases productive. But it wasn't mental or bodily comfort that Bushell needed: it was freedom, whatever the privations it cost (one of his worst experiences, he would relate, was contracting piles from using a dock leaf while on the run) and the chance to continue fighting the enemy.

Stalag Luft III, a large prison camp at Sagan, eighty miles east of Berlin in what is now Poland and far from any neutral country, was an especially good camp and had only been opened the previous spring, initially for captured officers of the RAF and Fleet Air Arm. Bushell was soon joined by 'Paddy' Byrne, of Barking Creek fame, and Aidan Crawley. The north compound to which Bushell was committed could almost have been a luxury camp; it held a thousand prisoners, was spacious, and boasted private kitchens and washrooms with every barrack. There were excellent facilities for entertainment and the well-intentioned commandant Von Lindeiner hoped the British prisoners would enjoy their stay and even wish to remain in Germany after the war. The north compound had a library, a theatre with shows that were put on by the prisoners, a camp radio station, a camp newspaper, and a wide range of sports, schools, and athletic activities of very kind. The camp was run by the Luftwaffe, and its guards tended to be sympathetic and kindly. Diet was meagre, but no worse than for German civilians, and it was supplemented by Red Cross parcels, which were fairly shared. The North compound where Bushell was housed was well organised, a small town unto itself; it was even likened to a holiday camp. It would have been possible to get by fairly comfortably by making the best of what the camp offered, taking the advantage of learning German, and staying out of trouble. Some did. None of this appealed to Bushell.

By concentrating all RAF officers in one compound the Luftwaffe had made a big mistake. They had created a vast pool of talent and discipline representing all professions and crafts, from doctors and engineers to plumbers and electricians and artists. All had survived a near-death experience, were volunteers, and understood the necessity of discipline in combat. All it took was organising skill, leadership, and the belief that still had a role to play in the war. To this army of prisoners MI9 smuggled wireless parts and escape material inside charity parcels, in contravention of the Geneva Convention. In return the prisoners systematically bribed and charmed the guards for outside information, a relatively easy process, which was sent back to London in coded messages home.

Not only the commandant but the senior British officer, Wings Day, had advised Bushell to take no further chances. First he had a spell in the 'cooler', the camp gaol, to undergo. This was so overcrowded with delinquent prisoners that those assigned to it had to wait their turn until a solitary-confinement cell was available.

Bushell emerged from the 'cooler' to the dismal news that, according to Pearson, his fiancée, Peggy Hamilton, in London, to whom he had given a diamond ring and directed all his service pay, had announced plans to marry a person of grander social standing. He had probably become prepared for such an outcome because her letters had dropped off, and now he told his solicitor in London to do what he could to recover the money though, he admitted, it might not be possible.

The physical effect of two years' privation and harsh treatment capped by solitary confinement was apparent. Stanford Tuck, Bushell's wingman with No. 92 Squadron earlier on the day he was shot down, arrived at the camp after having also been shot down. Tuck was now an acknowledged ace and a wing commander with the DSO, having leapfrogged over Bushell in rank. The two-year time lapse emphasised the stark change in Bushell's demeanour: no longer amusingly mischievous but, with memories of victims' treatment by the Gestapo, burning with a hatred of the Nazis that matched the rage of Achilles. He flung himself with such intensity into the theory and practice of escape that, after playing minor roles in several escape bids, he rose rapidly through the posts of ascending seniority in the escape organisation to intelligence officer, and finally to its top position – chief executive or 'Big X' of the North Compound.

As 'Big X', Bushell studied case histories and learned from past mistakes, the commonest of which having been their haphazard, independent, and uncoordinated nature. Success called for scale and imagination, and he introduced a new and important concept, collectivism. There were to be no more private enterprises but a concentration on a highly efficient centralised organisation. The scale of the new attempt was to be massive in scale and imagination, an enormous feat of engineering and improvisation. As a corollary, there were to be no more inflexible timetables, and if for any reason the guards' (or 'ferrets') suspicions were aroused, all work was to cease immediately and not resumed until the security department gave the all clear.

To support the master plan Bushell organised departments for tunnel construction, carpentry, forged documents, rations, intelligence, and security. His nimble brain cut through to essentials. His single-mindedness drew from the secret briefings he had received from MI9; he knew his duty and embraced it. He inspired every member with a powerful will to escape, and when he summoned the committee, what he said stunned them. It was pure Roger Bushell. 'All of us', he said, 'are bloody lucky still to be alive. We have escaped from crashed aircraft or baled out of them, and we have been given a second chance. The war is still on, we are still in it; we are air force officers, and we mustn't waste this chance to continue fighting the Hun.' Even though few escapers could expect to get back home, he said, they would still be aiding the Allies by tying up German troops who would be diverted to finding them instead of being on the front line. Besides, some of them at least would likely make it. With this Bushell embarked on his third escape attempt fully aware of the grave personal risk he ran.

Three tunnels were to be constructed, and they were to be of such refinement that discovery of any one would lead to the belief that it must be the only one. To avoid a security leak the word 'tunnel' was banned from all discussion; they were to be called Tom, Dick and Harry. The prisoners bent their entire energies, diverted every useful item of food or material, subverted every sport or educational group, directed every imaginative talent, towards the objective of mass escape. The guards were less alert than the prisoners, who maintained an hourly and daily log of each guard's movements while constantly hatching escape plans.

Every newcomer had to be interviewed and vouched for by two prisoners to prevent infiltration by 'plants'. New arrivals at the compound were always impressed by their first encounter with Bushell, when he grilled them on what they had seen of the local area. His rather sinister appearance, with the gash over one eye, his forceful personality and well developed powers of interrogation, lent an awe-inspiring quality to the grim and clandestine surroundings of an improvised headquarters.

Bushell's intensity concealed magnanimity. 'Goon baiting', playing practical jokes on the guards and undermining their morale, was an understood responsibility of the prisoner, not just a game. Despite his mastery of the art, Bushell sometimes expressed remorse. 'It's not really fair,' he would say, 'some of these poor bastards are so simple they haven't a chance.'

Believing strongly in the value of high morale Bushell didn't forgo other camp activities, and he played the lead in a theatre group drama. Such harmless activities as acting and learning languages were also intended to convey the impression that he wasn't planning to escape, since the guards were watching him closely.

In a mock parliament he made a powerful and dignified speech as the Conservative prime minister, managing to identify Conservatism with the principles for which the war was being fought. Major John Dodge (nicknamed 'the Artful Dodger' in tribute to his many escapes) elaborated Bushell's speech until he was able to claim that Conservatism descended from Christianity. Bushell, trained to argue any case, then crossed the floor of the House and spoke for the Labour Party: 'The picture you have just painted of Christ as a capitalist simply appals me.'

The Germans were, of course, well aware of the centuries' old idea of escape tunnels, but they thought they had taken every precaution against them. The compounds were situated over soft sand that would collapse into a tunnel. The huts were raised over two feet from the ground so that guards could see any activity below the floors. Small microphones were placed underground to detect the sound of digging or scraping.

None of this, however, had prevented the Trojan Horse escape in 1943. Sport and exercise was such a common and behaviour that when a wooden vaulting horse was carried daily and returned after being used for exercise, it seemed innocent enough. But inside the horse were two prisoners, and the horse would always be placed over the same spot on the ground. The men inside would remove a loose board covered with dirt on the ground and start digging. The prisoners running and vaulting above blanketed the sound of the tunnelling from the microphones. When the horse was picked up and taken away with the tunnellers and their earth inside, the board was replaced over the hole and again covered with dirt. Three prisoners actually managed to escape when the tunnel was completed and even made it back to England.

Successes like this, though minor and rare, inspired the prisoners to keep trying. 'Tom' was discovered by sentries, and 'Dick' was then used solely as a repository for sand as work proceeded with 'Harry', now the only chance. There was an ingenious solution to the problem of visibility under the hut. The entrance to Harry was hidden by a trapdoor under a heating stove, a flexible stovepipe extension allowing the stove to be moved with wooden handles. While a vertical shaft under the stove was being dug a fake air duct concealed the workers and there was always a fire burning so that its heat would keep the guards at a distance. From open, the trapdoor could be closed and sealed and the stove moved back in twenty seconds. Earth was disposed of by several means, including long bags under the prisoners' trousers that could be emptied on the ground by pulling a string.

Food and escape equipment were provided by the organisation for over two hundred escapers, considered the most optimistic estimate for the number that would get through the tunnel before it was discovered. If everything worked perfectly it would be possible for one man to go through every two minutes, making a total of two hundred and fifty during the eight hours of darkness. Long experience had taught, however, that there would always be hitches.

'Harry' was an amazing achievement, and Bushell's accomplishment has been described as a supreme example of project management. No doubt he had learned a lot from starting up No. 92 Squadron. With a length of 336 feet, twenty-eight feet deep at the entrance in the north compound and twenty feet at the exit among trees outside the double electrified wiring, shored up by wooden boards torn from prisoners' beds, it was furnished with electric lighting, ventilation operated manually by air pumps, and relays of trolleys connecting at three 'half-way houses', enlarged underground spaces, to pull prone escapers singly to the far end through the tight, two-feet square passageway. The sandy walls of the tunnel were supported by wooden bed boards and other scraps of wood. 'Harry' had taken two hundred and fifty men working full time a year to dispose of the sand it displaced. A plan of highly coordinated teamwork was devised to dispatch the maximum number of men in the minimum time. Except for about forty priorities who were thought to have the best chance of reaching England, each man on the escape list got there by drawing from a hat. He had his belongings checked by the inspection committee to obviate jamming in the tunnel with excessively bulky packages, was given an allotted time to arrive at Hut 104, which housed the entrance, and was thoroughly indoctrinated in his drill. The scale of planned escape was as ambitious as the tunnel construction and the escape plan itself. Bushell's plan called for the escapers to be divided into two groups of a hundred according to their likelihood of success. The first group, called 'serial offenders', spoke German or had a record of escapes and were granted priority. This qualified Bushell on both counts.

The organisation fixed the night of 24 March 1944 as the one for the breakout, twelve months after commencement of work on 'Tom', 'Dick' and 'Harry'. Every known factor had been weighed: the weather would be suitable for travellers on foot ('walkers'); there would be no moon; a strong wind would disturb the adjacent pine forest and drown any sounds made by leaving the tunnel.

There was acute tension on the evening of the break, the climax of several weeks' accelerated activity. Aidan Crawley writes in *Escape from Germany*:

> Many of us felt that we had nearly come to the end of a job that had been our whole existence. We had spent more than a third of our conscious time down this hole out of the last ten weeks and some of us nearer half of our time; when we were not actually working all our thoughts seemed to be taken up by the tunnel's needs. Even in bed it was on our minds. We had lived 'Harry', slept 'Harry' and eaten 'Harry', for sand seemed invariably to be on one's hair, one's ears, one's eyes and somehow to find its way into one's food ... Oddly enough, it was with a feeling of sorrow that we went below to complete the final stages.

From midday the engineers finished off final details, connecting wiring and installing extra lights, while the forgery department filled in dates on the forged papers. A little after nine o'clock two engineers went to open the exit. Every man was in his place, and zero hour was nine-thirty. Then there occurred a train of mishaps: there was a delay in opening the shaft, and not until ten o'clock did those waiting down the shaft feel the gust of cool air which told them the surface had been broken. Word was then passed back that the exit, contrary to plan, was several yards short of the trees. As the papers were date-stamped Bushell decided that the escape must continue, and hurriedly conferred with his colleagues on the escape committee to work out a revised method of control at the exit, necessary to avoid detection by the guard in their lookout posts. As the escapers moved forward, each lying prone on his trolley and pulled by the man ahead, further delays were caused by those who had broken the baggage regulations and got stuck in the tunnel with the bulkiness of their suitcases. The rate of departure dropped from two to twelve minutes per man. To add to these complications, an unexpected air raid on Berlin caused the camp electricity to be switched off, and with it the tunnel lighting. Over half an hour was lost as margarine lamps were substituted.

Bushell was noticed to be calm but more thoughtful than usual. Dressed as a businessman he had teamed up with Lieutenant Scheidhauer of the free French Air Force, with whom he planned to travel by train to Alsace. Both were on the priority list, Bushell being number four. As delays began, Bushell, wearing a civilian suit and converted service overcoat with astrakhan collar, with a felt hat, a briefcase in his hand, glanced at his watch. His wit hadn't completely deserted him and he called down the shaft, 'Tell those buggers to get a move on; I've got a train to catch.'

Bushell and Scheidhauer caught their train at Sagan station. Two days later, during the most extensive search the Reich had ever been forced to mount for escaped PoWs, they were recaptured at Saarbrücken railway station by security policemen and taken to Lerchesflur gaol. There they were interrogated by the *Kriminalpolizei* and admitted being escapers from Stalag Luft III.

Seventy-six prisoners escaped from the north compound before a guard, straying by chance from his normal beat to walk along the fringe of the wood instead of beside the wire, accidentally discovered the tunnel mouth at 4.55 a.m. Three escapers reached England and the rest were recaptured; five of these were sent to the concentration camp at Sachsenhausen, three to the camp at Barth, and fifteen back to Sagan.

Bushell's plan to tie up German resources was hugely successful. The news of an incredible mass breakout was elevated to Hitler, who was incensed when Himmler told him, with calculated exaggeration, that it would take millions of hours by the *Landwacht* paramilitaries to round up the dispersed escapers. Hitler was also afraid of an uprising among the foreign workers. At a stormy meeting with Goering, Himmler, and Keitel, whose objections he overrode, he gave instructions for all the escaped prisoners to be shot. He later allowed that to be modified to most of the escapers, which was translated into fifty, who were to be shot, according to Gestapo instructions, 'while trying to escape, or because they offered resistance, so nothing can be proved later'.

The candidates for execution were selected apparently at random and there is no evidence that Bushell was singled out. Orders were received by teleprinter from Gestapo headquarters in Berlin. Bushell and Scheidhauer were handcuffed behind their backs and driven in a car along the autobahn leading to Kaiserslauten. The car was stopped after a few miles, the handcuffs removed and the prisoners ordered to get out and stand on the grass. They surely knew what was coming. Both were shot in the back, Scheidhauer dying instantly, Bushell after a few minutes and two bullets. It was 29 March 1944.

It appears that Bushell the lawyer accepted the Germans' right to shoot him as a spy while in Prague but not as an escaped PoW officer. In his 2013 book *The Great Escaper* Simon Pearson persuasively suggests that besides being closely connected to MI9 and organising a system of coded messages from the PoWs about such matters as the location of V1 sites, Bushell was implicated in the successful plot to assassinate Heydrich in Prague. All the circumstances point that way: his proximity to the event, the timing, and his close association with the known plotters. What Pearson more firmly concludes is that Bushell's downfall was due to being a maverick who challenged authority all his life and that at 'Stalag Luft III, he broke the rules once too often'. Yet Bushell did respect authority; he knew flying was impossible without it, and after leaving 601 to run No. 92 he assumed authority. But he admitted to excessive self-confidence and it might be truer to say that, from the ski slopes to writing off Max Aitken's Aeronca, to plunging into a circle of Me 110s, to associating with the Resistance, to spying for MI9, to a central role in the Great Escape itself despite serious warnings, it wasn't just authority Roger Bushell challenged, but fate itself.

Chapter 12

Victory and Bankruptcy

> Soldiers who wish to be a hero
> Are practically zero,
> But those who wish to be civilians,
> Jesus, there must be millions.
>
> Norman Rosten, *The Big Road*

Prime Minister Winston Churchill left the Potsdam Conference in Germany in August 1945 to fly to London and await the result of the first general election since 1935. He was at the peak of success and prestige and, with the death of Roosevelt, the second most popular person in the world after Stalin, a distinction of sorts. As leader of one of the three victorious nations, he was participating in decisions that must have been deeply satisfying. How was Germany to be punished? What should the post-war order be? How was future war to be prevented? These were different by far from the decisions of life and death with which he had wrestled in 1940 under a stress that few men could have borne when he told the British people, without subtlety or reservation, that 'blood, toil, tears and sweat' were all he had to offer; that they stood alone because 'Hitler knows that he must break us in this island, or lose the war.' Hitler was now cinders outside a deep underground bunker in Berlin, which the Russians now occupied, or in Moscow. Churchill had been the inspiration and moral force that had sustained the public's spirits and confidence in pursuing a war, in the face of political opposition, that by all logic couldn't be won. He was not only the prime minister but the war leader and hero who on 8 May 1944 stood on the balcony of Buckingham Palace alongside the royal family acknowledging the adulation of the crowd.

Yet Churchill lost the election, and the morning after all votes had been cast he knew it, waking up abruptly with a stabbing pain and the sudden, cold fear that he had lost. His Conservative party was thrown out of office in a landslide and, now merely the leader of the opposition party, he never returned to Potsdam. Overnight he had no power. His place was taken by Clement Attlee, the quiet but competent deputy prime minister, now prime minister and leader of the newly victorious Labour Party. It has been said that the election was not a repudiation of Churchill himself, since he retained his MP seat, but of his Conservative party. But it was. His Epping constituency was a safe seat that he couldn't have lost if he had tried, and nationally Churchill, by his own choice, *was* the party. Yes, he had been wonderful back then, inspiring. But that was then. His geopolitical senses were as acute as ever and he was quick to perceive the new Soviet threat to world peace. But he had never tried to understand the British public's mind; rather than follow it, he had become used to shaping it. He thought at least the armed forces would support him but he was wrong about that too; most servicemen hadn't been fighting for the Empire but for the country, and not for the pre-war order but for a better one. His BBC broadcast during the election campaign was an embarrassment. He lost all credibility by likening a possible Labour government, headed by no less than his own former deputy prime minister, Clement Attlee, and other loyal members of the coalition cabinet, to the Gestapo.

The surface tranquillity of the British coalition government under Churchill's leadership throughout war had hidden powerful undercurrents. The Cabinet had been disappointed by his refusal to consider the matter of social reform after the war, a clear demand of the civilian armed forces. Many aircrew, including officers, wondered why sergeants and officers with the same skills and courage as they should be treated differently. Young officer types who had been trained to fly by experienced sergeant pilots who called them 'sir' wondered why that was. Was it the school they went to? Their connections? Their accent? The way they held a knife and fork? Their taste in food and drink? Ned Grosvenor would have had no hesitation in saying 'yes' to all of the above.

Desperation had driven the need to pull in the lower classes for the supreme war effort, while deaths opened up pathways for rapid promotion. Ground crew corporals had been trained as sergeant pilots and sergeant pilots had been made officers and moved up through the ranks to high command, as did John Bisdee, who began the war as a VR sergeant pilot and ended it as group captain with a DFC and OBE. Twenty-two-year-old Gilbert Henry Harnden joined 601 at Hendon in 1933 as an aircraftman, stayed with the squadron as a flight sergeant throughout the Battle of Britain, North Africa, and Italy, and ended the war as a flight lieutenant at wing headquarters. Aircraftman Second Class Jenson zoomed from Brian Thynne's metal rigger in the 1930s to group captain, exposing the human potential that class barriers had suppressed. There was now a widespread public determination to make these changes permanent, supported and implemented by the new socialist government. (Within a few years all pilots would be commissioned.)

The other, formerly overlooked half of the British population had also undergone a dramatic change in self assessment. Again out of desperation, women had been drawn into manufacturing, farming, and ferrying aircraft from factories and assembly centres to RAF bases with the ATA. Even the future Queen Elizabeth had been a mechanic and ambulance driver. Class was no longer admirable or amusing. As writer J. B. Priestley put it, the country had been bombed and burned into democracy. Millionaires, while not detested, were no longer subjects of amused admiration, although the word was still attached, whimsically, to No. 601 Squadron.

Classes still remained, but Britain's old class system had crashed in flames. It had served the country well in its time, but no longer could. Titles remained too, but increasingly were earned by accomplishments of national advantage or charitable works, and they no longer necessarily signified authority or entitlement. Neither were they necessarily correlated with wealth: taxes on incomes and estates took care of much of that.

Post-war Britain was nothing like pre-war. Physically, that was obvious. London was piled with rubble and dust, and in Graham Greene's ironic words 'pleasant with all those open spaces and the rather Mexican effect of ruined churches'. Food, clothing, and soap were rationed. Cars and luxuries were for export only, the telephone service was wait-listed with the General Post Office. Dock workers were striking. Everywhere was drab. The euphoria surrounding victory in Europe and the return home of the armed forces rapidly yielded to everyday miseries and disinterest in America's continuing war against Japan. The war had bankrupted Britain while enriching America, its mortgager. People wondered where the fruits of victory were. One retired 601 ground officer complained, only half joking, that he had 'expected to have a full time German servant after the war'.

Though millionaires in No. 601 had dwindled to zero, Grosvenor's Legionnaires had been replaced by true foreign legionnaires from around the world, and his Flying Sword had carried on. Britain's war was others' war too, a measure of the wrath Germany had drawn down upon itself. Lost blood had been replenished by the Empire Air Training Scheme; now there was no empire. Neither were there heroes. RAF bomber pilots were parodied in press and radio as moustachioed types with polka-dotted scarves saying 'Wizard prang' and 'Good show!' while people had tired of motion pictures in which Spitfire pilots – always Spitfires – with cultivated accents called out 'Tally-ho', 'I see them, Johnny', and 'Over to you, over'. The Battle of Britain was remembered as a great event, though doubters were beginning to question its importance or necessity. The Battle of Malta, always remote in British minds, had already sunk below the subconscious.

The war amnesia wasn't confined to Britain. Bill Matheson regretted in his book, *A Mallee Kid with the Flying Sword*, that a well-known Australian author seemed to 'find it difficult to understand why many thousands of young Australian men volunteered to help fight a war on the other side of the world. Perhaps today we live in a more cynical and self centred world. It may have been different sixty years ago.'

Churchill's eloquent 'so much owed by so many to so few' speech during the Battle of Britain resounded around the world and became almost a cliché. Less well remembered are his preceding words: 'The gratitude of every home in our Island, in our Empire, and indeed throughout the world, except in the abodes of the guilty, goes out to the British airmen who, undaunted by odds, unwearied in their constant challenge and mortal danger, are turning the tide of the World War by their prowess and by their devotion.' No fault can be found with this. The squadron's adversaries may have been just as brave and patriotic, but they were the aggressors and we owe them nothing.

Often it was those who had lost the most who best understood what the suffering had been for. After the loss of her husband Michael, Carol Doulton wrote to her father in America, 'I can't dictate to other people how they must feel, but to me it is wrong and wicked to weep and say all the silly things about why, and "he was so young" etc. Anyone who knows Hitler knows why, and what better way to die. It was Michael's way of dying, and it is a fine way.'

Pride in the Legion and its pilots ran deep. An airman, Harold Brown, who joined 601 at Tangmere, wrote, 'It was not long before the Battle of Britain was in full swing. The next few months became very hectic, at first the pilots were doing two days' flying and one day's rest, but when things really intensified the rest day was forgotten, they were even afraid to go to the loo in case they were scrambled (get in the air immediately), before they had finished their business.'

For him there was no begrudging the rich and privileged; they had clearly been paying their dues. After noting the succession of deaths, the strain imposed on these young men, Brown added,

> One had to admire these so-called millionaire playboys. They could have still been in civvy street instead of fighting for their lives. There are many

> stories of this period already well documented, I will only mention one as it is, in my opinion, very special. Billy Fiske, a very rich young American, (and a gold medal Olympian) joined the squadron. He could have been living a life of luxury in America, but decided he had to do his bit in the fight for freedom. Sadly he didn't last long. His aircraft was badly damaged in a dogfight, he crash landed on returning to base and died from his injuries the following day, the first American to die in World War Two. I don't know why, but despite coming from a very rich family, his grave at Tangmere remained untended and became overgrown. Just recently the Old Comrades Association of the squadron did some fund raising and now the grave is a fitting memorial once again.

From all parts of the world men who had been with the Legion at the outbreak of war returned with high rank and rows of decorations to begin life afresh. They were the lucky fifty per cent.

Of the twenty-four pilots with No. 601 Squadron in September 1939, fewer of whom were still operational by time fighting began, twelve had lost their lives – four over Dunkirk, four in the Battle of Britain, three in flying accidents or subsequent operations, and one murdered by the Gestapo.

Sir Nigel Norman, Bt, CBE, Sassoon's successor as 601's CO in 1931, died in an aeroplane crash on 19 May 1943 while returning to his duties as Sir Arthur Tedder's adviser in North Africa on airborne warfare and inter-service coordination. After serving in France during the early part of the war he was posted as wing commander in charge of the central landing establishment at Ringway, Manchester, which airfield his firm of Norman and Dawbarn had helped build. He became aware of a need for coordination between the three services and a new attitude to airborne warfare and its potential value, which was highly successful under Tedder. Together with Squadron Leader Louis Strange, a redoubtable aviator with the rare distinction of a DFC from both world wars, Norman established a force that became famous and proved itself on many occasions, despite a sorry failure at Arnhem caused by delay and poor intelligence. While at Ringway, Norman was issued with a light aeroplane for personal use. It happened to be the very Leopard Moth he had been forced to sell to the government for £700 when the war began and all private aircraft were requisitioned. After the North African landings in 1943 he joined Tedder's DAF. A series of important consultations prior to the Sicily landings brought him back to England, and he was returning to Africa in a Hudson with the pilot and some members of his staff when trouble forced them to return to the airfield at St Mawgan. Soon after taking off again one of the motors cut, and Norman left the co-pilot's seat to warn the passengers to fasten their straps and brace themselves for the inevitable crash, having only time to fling himself on to the bed that had been made up in his cabin with the words, 'This is my only chance!' The aircraft crashed and burned out. Norman and one passenger were killed.

In the summer of 1941 Wing Commander Brian Thynne, now a controller, sat impatiently in his underground office at Fighter Command headquarters in Stanmore Park, Middlesex. He had had the significant job of Sector Commander of No. 13 Group in the north of England during the Battle of Britain, helping to direct Dowding's brilliantly successful operations over Newcastle when on 15 August 1940 the Luftwaffe launched its simultaneous attacks along the southern and eastern coasts of Britain. At such times the work had been exacting, vital and rewarding. Now, the raids over, it was tedious and unspectacular.

One consolation, Thynne discovered, was the controller's cloak of anonymity, coupled with a delegated authority, which suited his impish spirit as it enabled him to 'roar up AOCs and get things done'. On the other hand, matters were sometimes referred to him that neither his rank nor his knowledge equipped him to handle. When Bomber Command telephoned to say they had caught a spy trying to steal an aircraft, Thynne was quick to take off the cloak. With calculated indignation he said, 'I am not the Provost Marshal, neither am I in Bomber Command.'

But the caller was insistent. 'This man says Fighter will know him. Caught him about to take an Anson; strange chap, says his name's Belleville.' Thynne bolted upright. 'Is he tall with thinning fair hair?' 'Yes, sir.' 'Put him on the line,' he said.

Thynne relished the moment. 'Now listen,' he began, cutting short the new voice, 'do you understand English?'

'Of course I bloody well do.'

'Why, yes, of course, all spies speak English.'

Belleville began an angry explanation: he was a ferry pilot with the ATA; he had been instructed to pick up an Anson; he had forgotten his papers and nearly taken the wrong plane; no, he was not in uniform—

Again, Thynne cut him short. 'You had better reserve your defence, but at least you can be assured of a fair trial in this country.'

The explosion of invective this provoked was, Thynne considered, his brightest moment in 1941.

A pilot serving under Thynne at a later date won the DFC. When the award was gazetted the pilot confided with some anxiety that his name was false; he had been a pre-war member of the RAF but had been cashiered for a low-flying offence, only to rejoin under an assumed name on the outbreak of war. The DFC award with its attendant publicity threatened to expose him. As the consequences could be serious, Thynne took the best legal advice and summoned the pilot to his office. He explained that he would have to be proceeded against for false attestation, then delivered a sermon, at the same time pointing out unambiguously that no one could be charged twice for the same offence. When he concluded by asking sternly, but without the authority to do so, whether the officer would elect to go before a court martial or take his CO's punishment, there was no hesitation in the answer. Thynne gave him a reprimand and helped celebrate his DFC.

When the facts came to light the Air Ministry was not happy, but the ever-accommodating AOC, AVM Saul, was more tolerant. 'You've got away with murder, Brian,' he said, 'but for God's sake don't do anything like this again while you're in my Group.'

Whitney Straight, shortly after his escape from France and awaiting his next assignment, was cycling with his wife along a country lane in the south when an AA scout stopped him. 'Are you Mr Straight?' he asked. 'Well, sir, you're wanted at the Air Ministry.' Straight was given three choices of command: the fighter station of Hornchurch, a Spitfire wing in Malta, or a transport group in Cairo. He left the choice to the RAF and was given 216 Transport Group, Cairo, which rendered valuable service to the DAF from the time of Alamein on. He left the air force as an air commodore with the CBE and became deputy managing director of Rolls-Royce Ltd. Some years after the war he drove through France retracing his old escape route, seeking out the people who had helped him. At Belbec he found himself face to face with a farmer whose expression turned to fright as Straight described his crash-landing, his attempted firing of the plane and hurried escape, for he was the man who turned him away before alerting the Germans, and the penalties for collaboration were still severe. Air Commodore Whitney Willard Straight CBE, MC, DFC, ex-managing director and CEO of British Overseas Airways Corporation at age thirty-four, and ex-Deputy Chairman of Rolls-Royce, died in 1979 at the age of sixty-three.

Aidan Crawley spent his first night of captivity in Rommel's desert headquarters before being flown via Greece and Austria to spend four years in prison camps in Germany and Poland. At Schubin in Poland in 1943 he escaped by tunnel but was recaptured at the Swiss border after travelling though Germany by train. When they moved him to the notorious north compound of Stalag Luft III he became the head of Bushell's escape intelligence organisation. Being an excellent linguist he acted as interpreter for the senior British PoW officer when the new camp commandant broke the news that fifty RAF and Allied officers had been shot after the mass breakout. They had been shot, said the commandant, 'while resisting arrest or attempting further escape after arrest'. 'How many were wounded?' asked the British officer. The commandant replied, 'I am not authorised to answer questions or give any further information.' When pressed for an answer, he muttered with visible discomfort that he didn't believe any had been wounded.

When informed of the mass escape from Stalag Luft III, Hitler at first wanted all seventy-three recaptured prisoners shot, as a well as the camp commander, von Lindeiner, the camp's architect, and all the guards on duty at the time of the escape. Under pressure from the army chiefs who referred to the Geneva Convention, and probably fearing post-war retribution, he compromised by ordering Himmler to shoot 'more than half' of the escapers, which was translated into fifty. Roger Bushell was killed by Gestapo official Emil Schulz, who at his trial testified that he shot Bushell twice since he was writhing in agony after the first shot. Schulz's wife's written plea to the Queen was ignored and he was hanged.

Von Lindeiner was fired from his post as camp commander and narrowly escaped court martial and execution by feigning mental illness. He served as second in command of an infantry unit defending Berlin against the advancing Soviets, later surrendering to British forces, and testified at the inquiry into the mass shootings. Since his conduct had always been in accordance with the Geneva Convention and had won the respect of his prisoners, he was released from prison in 1947, dying in 1963 aged eighty-two. The Luftwaffe had nothing to do with the murders, and a later commander of Stalag Luft III, *Oberstleutnant* Erich Cordes, was so appalled that fifty escapers had been shot that he allowed the prisoners to build a memorial to them, which still stands, and even contributed to it. For many years after the war the memorial could not be seen because the camp was in Eastern Germany and under the control of the Russians, who had no interest in the escape. Now it can be.

After his arrest with Bushell, Jaroslav Zafouk was sent to the fortress at Colditz, survived the war, married, and emigrated with his wife via Britain to Canada, though sadly they later divorced.

George Ward, a Legionnaire from Sassoon's days, like Crawley entered politics but as a Conservative. With the subsequent change in government he succeeded Crawley as Under Secretary of State for Air.

In 1945, servicemen were counting their blessings, and in the next few months the Legion's members of all nationalities went home hoping to fit into the post-war pattern. After demobilisation Bill Matheson returned to life in the family farm in The Mallee, Australia. He married a local girl, Jean, and had four sons, one of whom now manages the farm. Matheson obtained and kept up a private pilot's licence and flew light aircraft until he was eighty, returning to London twice for No. 601 Squadron all-ranks reunions at the Victory Club. Proud of his service, he wrote and self published *A Mallee Kid with the Flying Sword* in 2005 at the urging of his family.

The war had been long, exhausting, and bloody for Britain. The next few years didn't promise to be happy ones. As men handed in their rifles and claimed their 'demob' suits, women joined the lengthening queues for the things that reunion required. As fast as memorials and cenotaphs to fallen RAF locals sprouted in towns and villages, not only in Britain but across the Channel and elsewhere, people sickened of war and of uniforms.

Nevertheless, a question was tabled in the Commons on the subject of reconstituting the Auxiliary Air Force and on 27 March 1946, the Under Secretary of State for Air, Mr John Strachey, replied that, 'Reserve Command of the Royal Air Force will be reestablished in the immediate future. Its function will be to train and maintain adequate reserves of flying and ground personnel …' So, for the time being, the Auxiliary units were bunched together with the RAFVR, University Air Squadrons and Air Training Corps in what came to be dismissively known as 'Reverse Command'. That this arrangement was unsatisfactory was soon realised, and by 1948 the Auxiliary units were integrated with Fighter Command.

Again, the driving force was economics. The Auxiliaries of 601 now comprised a few blue bloods but increasingly twenty-year-old men who had completed national service with the RAF and fitted comfortably into the post-war notion of a democratised, part-time civilian defence force. Although the required eighteen months of military service in one of the three forces was a dreary interruption in many recruits' lives, for those who craved and were selected for pilot training it was a gift from the fairy godmother, even with the additional three months' jet conversion required for those joining an Auxiliary squadron.

The Air Ministry decided to revert as far as possible to the pre-war basis of organisation, but a plan to separate the force into bombers, day fighters and night fighters got nowhere. All the squadrons became day fighter. The question that had been pending in 1939, 'Is the Auxiliary Air Force worthwhile?' seemed now to have been answered by the squadrons' own war records, and Strachey added, 'We attach the utmost importance to these non-regular forces, and it may well be that in the future it will be desirable, and possible, to develop them to a much greater degree than before the war.' Promising words indeed.

And so, against a sombre international black cloth Britain's new part-time air force was assembled. On 10 May 1946 the AAF was re-formed with the elevated title *Royal* Auxiliary Air Force (RAuxAF) and No. 601 Squadron began to recruit personnel at Hendon in June. The old territorial connections were restored, airfields allocated and commanding officers appointed. Squadron Leader the Hon. Max Aitken, DSO, DFC, MP returned as the first post-war CO.

Aitken's career in the war had been distinguished. He was a pioneer of night-fighter operations while in command of No. 68 Squadron, which flew Blenheims and later Beaufighters, this being the first unit to receive the effective Mark VIII radar sets, which enabled aircraft to detect night intruders with considerable accuracy and to attain a high level of success. In 1943 he commanded the Fighter Tactics branch at Headquarters, Eastern Mediterranean, and in 1944 operations in the Aegean. Later that year he returned to Britain to lead the biggest strike wing that existed, consisting of 138 Mosquitoes, from Banff against shipping in Norwegian waters. Having flown on the first day of the war, he capped his operational record with an anti-shipping strike off Denmark on the last. He became chairman of the family-owned Beaverbrook Newspapers Ltd, and entered Parliament. With Portsmouth sea pilot John Coote he established the Cowes-Torquay Offshore Powerboat Race. On inheriting his father's title Aitken promptly renounced it, saying, 'There can be only one Lord Beaverbrook.' At the first post-war general election Aitken was elected Conservative Member of Parliament, and the Legion appeared to be driving a wedge into Westminster. He was not the only man willing to drop rank to serve 601, being joined by three other former group captains and two former wing commanders.

One of the group captains, Sir Hugh Spencer Lisle Dundas CBE DSO and Bar DFC, nicknamed 'Cocky', had joined the Auxiliaries straight from Stowe School in 1939. Shot down twice and wounded in the Battle of Britain, he flew in Bader's wing from Tangmere in 1941 and proved to be an outstanding fighter pilot. He became one of the youngest group captains in the RAF at the age of twenty-four through exceptional virtuosity as a fighter leader, commanding 244 Wing in Italy of which 601 was a part. After the war he took a job as air correspondent for Aitken's *Daily Express* and cheerfully joined 601 as a flying officer, the second lowest rank and four levels below group captain, later to succeed Aitken as CO.

Dundas's successor, former Wing Commander Paul H. M. Richey DFC, was shot down three times in 1940 air battles and wounded. By war's end he had claimed ten victories. He was famous for his 1941 autobiographical book *Fighter Pilot*, widely regarded as a wartime aviation classic. He commanded the squadron during the three-month call up in 1950 but went abroad and was succeeded by Squadron Leader Christopher (Chris) McCarthy-Jones, a sales manager in business life, who was shot down and captured and became one of Bushell's escape organisation in Stalag Luft III, helping to make maps. McCarthy-Jones gave the squadron five years of stable stewardship and won a diplomatic victory in having Prince Philip, the Duke of Edinburgh, consent to become Honorary Air Commodore, replacing Sir Sholto Douglas (one of Dowding's former detractors and his successor), since several other squadrons had the same idea. Prince Philip showed a genuine interest in 601, dined with the officers in town headquarters, and visited the crew room at North Weald.

There was no shortage of pilot applications, and in a very short time the astonishing number of over four hundred applications engaged a selection team of four presided over by the CO, who imposed a minimum qualification of fifteen hundred flying hours. The result was a squadron of unrivalled experience in the RAuxF. Even the first four NCO pilots consisted of a former squadron leader and three former flight lieutenants. (The last NCO pilots in 601 were commissioned in 1954.)

At a time when most non-professional officers couldn't wait to sever connections with the armed services this was a measure of the pull that flying and squadron life exerted. A good example was Flying Officer Gordon ('little e') Hughes, so nicknamed because of his superstition about the letter 'E' which made him write even his middle initial in lower case. While a wing commander in the RAF photographic reconnaissance unit he carried out dangerous high-level, long-range unarmed flights in blue-coloured Spitfires and Mosquitoes, choosing this mode of operation because he was, if not a pacifist, averse to killing. Being averse to violence didn't guarantee immunity from it, and Hughes was shot down by flak over the Ruhr three days before the end of the war, spent nine months in hospital, and ended the war with a DFC and Bar, plus a 60 per cent disability pension, which he continued to receive during his service with 601.

The amateur, part-time character of 601 had been replaced by the war; now it began to flow back. At first the only source of pilots was decorated veterans, luminaries who took major demotions to fit into the small squadron flying complement led by a squadron leader. A few of the new pilots, for reasons it was hard to understand, were sergeants until all were commissioned in the early 1950s. There was a need, soon filled, for staff officers in engineering, equipment, medical, and administrative functions. A regular training officer and an adjutant, selected for both technical and diplomatic skills, were attached. The flying hours requirement was relaxed as national service pilots, each with 250 flying hours, fifty of them on jets, flowed in to make up fighting strength.

There was little difficulty in recruiting the ground crews, although it took longer. The squadron bought £100-worth of advertising space in London's tube trains and soon had enough applications to make up establishment. Among the airmen were bus drivers, policemen, a baker, a barrow boy, an architectural ironmonger, and a lighthouse keeper who would disappear for two months at a time between visits to Hendon where he would train intensively as a fitter.

Another application was the almost heart rending case of Mr Dunkle. He had been a flying officer on 601's ground staff in the North African desert, and had lost both legs in an accident. The only vacancy open was as a corporal in motor transport, which Mr Dunkle was eager to accept. At this point Air Ministry intervened and ruled that a legless man would be useless as a driver. In vain the squadron protested that their new sister squadron at Hendon, No. 604 Squadron, had just enrolled as a pilot Colin Hodgkinson, a legless wartime pilot who, now released from a German prison camp, couldn't wait to fly again. Though legless Hodgkinson was accepted as a pilot, legless Dunkle was turned down as a driver for motor transport.

On 10 July 1946 the first aircraft, a yellow Harvard FX 387, flew in, and the Flying Sword rose over the misty air at Hendon where it had last been when the war broke out. Towards the end of the year Spitfire 14s arrived, and Aitken was able to lead a fly-past for the AOC while its sister squadron, No. 604, failed to put a single aircraft in the air. The old spirit was back.

Max Aitken, under pressure of work with the Beaverbrook Press and as an MP, gradually handed over to 'Cocky' Dundas, who assumed formal command in June of 1948. Dundas's aviation articles for the *Express*, though informed, were characteristically forthright and provocative. The orderly room forgot to forward him a letter, which arrived through official channels seeking to prevent the publication of one of his articles, and it was published.

Air Chief Marshal Lord Tedder, now CAS, visited the squadron's cocktail party a few days later; the mess again looking exactly as it had in the 1930s with the china, crystal and silver reclaimed from Lloyds Bank. He left the anteroom with Dundas for half an hour's earnest discussion in the hall. Dundas returned to the party all smiles. 'We've reached an amicable agreement,' he explained. 'I'm going to carry on writing articles, and Tedder's going to carry on running the air force!' Lord Tedder left his uniform cap at the party, and sent his chauffeur back for it. The cap was filled with coppers and buttons and given to the chauffeur with instructions to pass it with its contents to His Lordship. 'Leaving my cap behind,' wrote Tedder later to the squadron, 'is an old trick of mine, but never, not even from Air Training Corps, have I received so niggardly a collection as one shilling and threepence.'

A superb leader in the air and on the ground, Dundas captured the Grosvenor spirit with ease, made this spirit the yardstick by which all activity was judged, and used his considerable authority with economy. For a dinner at Manston during summer camp he had the squadron silver and mess ornaments transported from Kensington and it looked as though 601 was going to behave itself, but the dinner broke up with the explosion of powerful fireworks under the table and reached pandemonium when the room filled with smoke as rockets impaled themselves in the ceiling and showered the diners with sparks. Before the entrée was finished not a soul sat at table and servants dared not enter the room. A notice, 'QUICK! QUICK! I'M ON FIRE!' was pinned to the Mayor of Margate's back, and he was drenched in soda water from powerful siphons. Air Marshal John Hawtrey, 601's long-ago adjutant, tried to cheer up the station commander: 'Don't worry about a thing: if they burn down the mess they'll pay for it.' Though the mess didn't burn, the damage came to £70, including the cost of a bicycle, which was ridden round and dismantled. The Legion, as always, paid.

Less famous than these COs were some highly experienced veterans. Flight Lieutenant R. G. 'Bob' Large, DFC, *Legion d'honneur*, had been a member of No. 161 Special Operations Squadron flying unarmed Westland Lysanders clandestinely at night, deep into France to deliver or recover agents of the Resistance. The procedures for pilot and passenger were highly specialised and required coordination and practice. Usually refuelling at Tangmere before crossing the Channel, Large had to rely on a combination of dead reckoning – calculations based on speed and wind – and by following rivers visually as these were usually reflective at night. The Lysander was a slow, high-wing monoplane with fixed undercarriage and an ungainly stance, helpless if attacked by a fighter but capable of extremely slow flight well suited for landing in small fields. One of Large's charges was Violette Szabo, a beautiful Paris-born Anglo-French woman, raised in England and a British agent, whose life was to end with capture, torture at Gestapo headquarters in Paris, and death by SS firing squad in Ravensbrück at age twenty-three. Large was always jocular and one of the squadron's best raconteurs, endlessly amusing and endlessly talkative. His dog Patrick was highly intelligent and obedient and Large's constant companion, even on some occasions in the cockpit of a Vampire.

Flying Officer Prince Emanuel Galitzine was among the upper crust drawn to the squadron. Born in 1918, he was the great-grandson of Emperor Paul I of Russia and fled with his family to London to escape the Bolshevik purges. He had always wanted to fly, and when the Russians attacked Finland he joined the Finnish air force to fight the Bolsheviks who had dispossessed his family. After Finland's collapse he returned to England by sea via Boston, Massachusetts, and Scotland. He joined the RAF, was posted to an experimental Special Service Flight in Northolt and credited with the highest-ever interception when in a pressurised Spitfire he attacked a similarly pressurised Ju 86P at the unprecedented altitude of forty to fifty thousand feet, his guns unfortunately freezing so that he couldn't score a kill.

Dr Nevil Leyton was appointed 601's official doctor. Nobody quite understood the motivation of this dapper and famous Harley Street headache and migraine specialist; he disliked flying and his eardrums were once pierced on a flight as a passenger to Malta in an unpressurised Meteor 7. Annual summer camps depressed him because he couldn't stop worrying about the fees he was missing.

The Spitfires, beloved but obsolete, were replaced by somewhat quirky de Havilland Vampire 3s in 1950, later 5s. They could fly faster and higher than the Spitfire and were easier to service but not much else. A curious shape befitting its name, but without its bite, its wings sprouted from a pod containing the DH Goblin engine, four 20 mm cannons – only two of which were ever fired and their recoil seemed almost to stop the aircraft in the air – and the pilot. There was an air intake for the engine in each forward wing root. The horizontal tailplane was attached to the wings by a narrow boom on either side of the pod. The pod itself sat very low and pointed upward, so that on start-up an airman had to stand by with an extinguisher if there was grass around to put out the fire from the initial spurt of flame. There could be no serious thought of escape from this machine and no such instruction was ever offered, but then jet engines were not supposed to fail – unless they ingested a bird, or the flame just went out, for which unlikely event there was no solution. The Air Ministry lived down to its reputation for putting hardware before human assets. Flying the Vampire could be unsettling at first. Lacking the grip of a propeller there was no immediate push in the back on take-off and it seemed the plane would never leave the ground. Unlike the Spitfire and Hurricane there was no bulky engine in front blocking out vision, which was helpful, and no swing on take-off from propeller torque, which although this too was helpful, somewhat diminished the satisfaction of taking off. Close formation was a little tricky at first because, as with all jets, engine response to movements of the throttle were delayed without the immediate pull or drag of a propeller, and one had to anticipate more. Like all jets it seemed heartless and less a steed than a machine. But, and this seemed significant, the squadron had moved into The Jet Age.

Once again, only the best would do. Town Headquarters resumed its place with lectures and dinner once a week. The portrait of Grosvenor once more hung over a crackling fire in the anteroom fireplace, and just below it on the mantelpiece the clock donated by Dickie Shaw. Officers began arriving from work on Thursday evenings and gathered in the softly lit green and gold anteroom for drinks before dinner, an assembly of vast experience spanning the breadth and duration of the war. Dinner was prepared by the caretaker's wife and served by him in the dining room. The silver, restored and gleaming and including handsome models of a Hart, Gladiator, Blenheim, Hurricane, and the Legionnaire charging with the flying sword, lined the centre of the long, candle-lit table. A typical dinner, meagre by the standards of Grosvenor's days but liberal enough for the time, would include canned soup followed by a lamb chop or a sausage – meat was rationed until 1954 – with claret passed around, followed by apple pie, port, and for hangers-on cognac.

As the old guard began to drop out the three sons of Sir Nigel Norman, Sassoon's successor as CO in 1931, trickled into the squadron in age order: Sir Mark, Desmond, and Torquil, all Etonians and well over six feet tall. The newcomers were for the most part little aware of this generational link with the past or with the squadron's legacy but readily absorbed its traditions, including red socks and scarlet tunic lining, blue ties, and the name Legion.

Few of the newcomers were upper class, but there was a hint of class nevertheless. Torquil Norman writes in his semi-autobiographical book, *Kick the Tyres, Light the Fires* that 'because of 601's history of flying to the rescue of Malta during the war we were treated like kings when we arrived for our summer camp and royally entertained by many of the local families'. Maybe so, but most officers and men experienced no such treatment. Not that they minded one bit. The upper layer sprinkled the squadron with a lustre that covered all its members, including the ground crew. In the mess, the crew room, Town Headquarters and in the air – save for assigned formation positions – all were equal. That on Sunday evening some would travel back to London for oysters at Bentley's before retreating to their homes or country estates while others found their way back to bedsitters or suburban homes didn't matter. When one of the first national service pilots expressed dismay at the Town Headquarters' mess bill of a fixed four guineas a month excluding drinks, 'the cost of a new pair of shoes' he said, the CO Paul Richey promptly halved the mess fee for all national service officers. This was now a millionaires' squadron in name only.

When the squadron's possessions were unpacked the pre-war Line Book was found and caused much merriment. But the names in it rang few bells. Sure, 'Ned' Grosvenor was the founder whose portrait hung over the fireplace at Town Headquarters; Max Aitken was a post-war CO and a popular newspaper icon; Willie Rhodes-Moorehouse was no doubt related to a dimly remembered First World War hero; Dickie Shaw was the pre-war member who donated the clock on the mantelpiece; and Nigel Norman was vaguely understood to have been the Norman brothers' father. But the others: Brian Thynne, 'Little Billy' Clyde, Roger Bushell, Rupert Belleville, Michael Peacock, Tony Knebworth, Michael Robinson, Loel Guinness, Jack or Huseph Riddle, Guy Branch, Ray Davis, Dick Demetriadi, Archie Hope, even Philip Sassoon, were just names.

The discovery of the Line Book prompted the start of a new one. The tone was familiar:

> 'I have completely forgotten what it is to like to sit in the cockpit of an aeroplane without a hangover.' Desmond Norman
>
> 'You know, that's the first dining-in night I've been to. We didn't have them during the war.' John Bryant (Typhoon pilot)
>
> 'Being a Knight of Malta, I can hardly chase the frip around in Valetta. Have to leave it to you younger chaps.' The (very) Honourable Peter Vanneck
>
> 'I must avoid the women in Malta.' Clive Axford
>
> 'They used to call me "Fly"; I pretended it was because I was aircrew.' F. L. Alf Button (Equipment Officer)

Harold Harmer: 'Nick [Nicholson] knows a drink when he sees one.'
Len Brett: 'That's about the limit of his knowledge.'

Ken Askins: 'Sorry I'm late, I've just been to an agricultural show.'
Peter Edelelston: 'Did you win anything?'

'Oh, it's alright. You have to, er, be a bit of a diplomat and forget a few old ideas.' Jock Agnew on the phone, a week after arriving as the new regular training officer.

Under the rationalisation of airfields 601 moved to North Weald, Essex, a famous Battle of Britain aerodrome. During the 1951 Korean crisis the Auxiliaries were called up for three months' full-time service.

Each year as a test of mobility and air gunnery the squadron flew abroad for two weeks' summer camp. Four times summer camp was held at Takali, Malta GC, where clear skies guaranteed uninterrupted air-to-air firing over the sea. Takali was five miles to the west of Luqa, no longer connected by the Safi Strip, and only a burned-out Spitfire, left around for the airmen's children to play on, and Denis Barnham's Flying Sword painted on the wall of the Naxxar Palace, recalled the far-away days of privation and struggle. To the land-locked airmen from England sunny Malta seemed truly exotic, but the naval types who lived there called it 'a goat-infested lump of bath brick'.

In 1955 there was a visit to Takali by 11 Group AOC the Earl of Bandon, popularly known throughout the RAF as 'The Abandoned Earl', arriving as passenger in a two-seat Meteor 7. That evening in a ballroom taken over for an all-ranks party the Earl startled everyone with a bravura performance. He undertook to balance a glass of beer on his forehead and then drink its contents without using his hands, and proceeded to do so without even removing his suit jacket or tie. Having placed the glass on his forehead while looking fixedly at the ceiling the Earl lowered himself to a kneeling position, first on one knee and then the other, slowly reclined until he was lying face up, the glass undisturbed, then by raising his knees to his head clasped the glass between them and continued rolling backwards, keeping the glass upright, until he was able to release it on the floor. The easy part now was turning round, gripping the glass between his teeth, standing up, and tilting his head back to drink.

When in 1955 summer camp was held at the ex-Luftwaffe fighter base of Wunstorf in Germany, 601 got a glimpse of how the other side had lived, and it was an eye-opener. The airfield had pathways leading through dense tall pine trees to the pens where Me 109s and Fw 190s had been shielded from sight while being armed and refuelled. The officers' mess was a dream compared with anything seen in the RAF, boasting a skittles alley in the basement and, intriguingly, a vomitorium in the lavatory, with leather-padded headrests, handles with push-button flushes and, presumably as a stimulus, coloured images of retching storks decorating the facing tiles.

In mid-1952 the Vampire 5s were traded in for the twin-engined Gloster Meteor 4, later Mark 8s. The Meteor was a robust four 20 mm cannon gun platform, standing high and with hand holds requiring something like cliff climbing to reach the cockpit. It had so much power that after take-off it could accelerate while climbing, thanks to two Rolls-Royce Derwent jet engines. It was capable of asymmetric flight, given enough speed and sustained leg force on the live-side rudder pedal, so that, as the old adage went, even if one engine went out the other one could be relied on to take you safely to the scene of the crash. With the Meteors came ejector seats, crash helmets ('bone domes'), and zippered flying overalls. The Air Ministry loved the Auxiliaries after all.

George Farley, the adjutant, made a marvellous discovery which he revealed at a Thursday evening meeting – an order, issued just after the war and long forgotten, permitted RAF aircraft to fly at low level anywhere in Western Europe, given twenty-four hours' advance notice. Low level is an excellent exercise in map reading and tactical formation, but it is also tremendous fun. Low level meant five hundred feet, but who could tell? By luck the Meteors had been fitted with long-range tanks in preparation for an annual camp at Malta. The pilots saw the doors to Paradise swing open and that weekend they streaked out in pairs across the brilliant tulip fields of Holland, scurried around the Arc de Triomphe, the Brussels Atomium, and Cologne cathedral, and buzzed ships in the Channel both going out and coming back. Group's alarming call to the squadron was as quick as it was predictable: 'What the hell are your people up to?' Dickie Smerdon, replacing Farley, referred to the order and quoted its reference. There was silence, but the following Thursday Smerdon advised everybody to act quickly because this weekend would doubtless be the last chance. It was. After the weekend, Group's telephone switchboard lit up like a Christmas tree. Calls poured in and the heavenly gates slammed shut.

The Norman brothers had been steeped in aviation culture their whole lives and all had flying experience before their national service. Desmond, the second son, was one of the few people to do an outside loop in a Meteor. He landed a Tiger Moth in the dead of a winter night at North Weald while the airfield was closed, turning over the Epping high street traffic lights and reading his primitive instruments with a stormproof cigarette lighter. He had his father's entrepreneurial drive, and while active with 601 was co-founder at twenty-six with his friend John Britten of the crop-spraying firm of Britten-Norman Limited, on the Isle of Wight, designing and manufacturing aircraft, hovercraft, and racing yachts. The B-N Islander, a simple, fixed undercarriage, twin-engine utility aircraft with high wings for easy loading and unloading, designed to replace the world's aging Douglas DC-3s for short hauls and island hopping, was spectacularly successful and is still in service around the world. For this success Desmond was awarded the CBE.

Desmond's piloting skill was exceptional and was accompanied by a high risk tolerance. He and Denis Shrosbree teamed up to perform impressive formation aerobatics at North Weald. They may have been pushing the boundary of risk too far when they attempted to emulate the brilliant Polish ex-RAF Battle of Britain pilot and former leader of 316 Polish fighter squadron, Janusz Zurakowski, whom they had seen give a stunning demonstration in a Meteor at the 1953 Farnborough Air Show. Stall turns of half a rotation are not difficult in a single engine propeller aircraft, but this was something else. At the air show Zurakowski's Meteor, a twin jet, soared vertically upwards until it lost flying speed and its nose probed the low-hanging clouds, where it hung for a moment, seemingly still, before cartwheeling through a complete rotation and a half, then plunging vertically down the same trajectory as its climb. Norman and Shrosbree, familiar with the type, believed Zurakowski must have cut one engine while simultaneously kicking the rudder and holding the stick over to execute the cartwheel while at near-zero speed. Their surmise was reasonable and they could rehearse everything in their minds, but as they were to find out it is one thing to describe a high-wire act and quite another to perform one. They took a two-seater Meteor 7 up at Malta and found a clear space over the sea. Norman was in control in the front seat and started his climb. When speed had fallen off he closed down one engine, 'but Des only throttled one engine back, Zura cut the engine completely, which made all the difference,' Shrosbree recalls. Their Meteor slid backwards and then within seconds there was a blur of sea and sky and the aircraft

> pivoted sharply down and we proceeded to rush downwards towards the sea. At about two thousand feet we pulled out of the dive and started to climb away, which is where Des said to me, THANKS DENIS, I'LL

TAKE IT NOW! Throughout the whole time I had been hanging on to the top coaming like grim death and hadn't touched the controls once! Des, for once at a loss what to do next except hold on, felt the stick push powerfully forward and thought, good old Denis, he has control, and let go of the stick!

It probably made some difference that they were flying a weary two-seat squadron hack, which had a known tendency almost to roll on its back when its flaps were lowered, whereas Zurakowsky had a customized G-7-1 prototype. And, good as Norman and Shrosbree were, it may be less than fair to Zurakowski to say throttling back an engine instead of shutting it down (in fact he did not shut it down) made all the difference. Nevertheless, without even trying, the two had performed probably the first and last cart-wheeling snap-roll hammer-head stall in the entire air force, and if they could have reproduced it they might have been invited to demonstrate it at Farnborough themselves. But they never tried it again.

Shrosbree, known as Jet Head by his colleagues at HM Stationary Office where he worked in 1954, took off one Saturday afternoon in the CO's Meteor with its distinctive red and black striped tail to impress a friend playing rugby at Richmond. He was completely unaware that just across the river from the Richmond grounds was Twickenham Stadium, and that at that very moment the RAF was playing the Army at Rugby there. In Shrosbree's words, 'I climbed to 10,000 feet across London (which was allowed in those days). It was a bit hazy but I identified the football pitches alongside the river at Richmond and stuffed the nose down, made a bit of noise over the football games, heading west across the river and did a couple of rolls.'

Inverted flight was a favourite act of Shrosbree's. By trimming well forward he could so easily hold the angle and altitude while on his back that he could probably have flown cross-country like that. He returned to North Weald and then home for another drab week at the Stationery Office, blithely unaware that he had just lit the touchpaper of an enormous rocket that would soon be heading his way. It wasn't just that he had overflown the adjacent Twickenham sports arena 'at which the RAF had been playing the army at rugby, and lost', as he says, but that among the audience were most of the Air Council, including [Marshal of the Royal Air Force, the top dog] Sir John Slessor'. Just as Slessor was about to enter his car he turned towards the sound of an approaching aircraft and saw a Meteor streaking across the field at full at throttle and low level, inverted. Horrified and embarrassed he shouted to an equerry, 'GET THAT MAN!' bumping his head on the car's door frame at the same time. The rocket took off.

The Special Investigations Branch sprang into action and immediately impounded all records of aircraft movements in the United Kingdom. A wire was shot off from No. 11 Group to headquarters all stations:

> UNCLAS AO 167. LOW FLYING AIRCRAFT COMPLAINT. AN AIRCRAFT REPORTED OVER TWICKENHAM AT 1640Z ON 27 MARCH HAS BEEN IDENTIFIED AS A METEOR MARK EIGHT. STATIONS ARE TO MAKE INVESTIGATIONS AND REPORT IMMEDIATELY BY SIGNAL. NIL RETURNS ARE REQUIRED.

This was serious stuff. Many pilots had been thrown out of the force for less and Shrosbree expected the worst. He shakily read the charge sheet when it came down, with its chilling introduction that seemed to portend a beheading: 'Whilst subject to Air Force law and being the pilot of one of Her Majesty's aircraft …' A controlled area, low flying, reckless endangerment, disobeying standing regulations, a lost rugby match, raising a bump on the CAS's head; even a team of Roger Bushell and Michael Peacock could hardly have got him out of this one. He saw his hopes of becoming an airline pilot circling the drain. His written report was a comical evasion:

> On the afternoon of the 27th of March, 1954, I was briefed for an aerobatic sortie on the local flying area, and set out from North Weald on a south westerly course. When I had estimated that I was clear of the London Control Zone, I descended in order to determine my position, as the visibility was very poor. I pinpointed myself at Kingston in error, and commenced to climb again, executing a roll at the same time. I didn't realise that I had been near Twickenham until recent enquiries were made and I apologise for violating Air Traffic Regulations in this manner.

Shrosbree's tortured explanation that after having inaccurately pinpointed his position in very poor visibility he just happened to perform a roll while climbing away must have prompted some snickers among the staff at Fighter Command HQ. The CAS didn't buy it. Shrosbree was summoned before AVM H. L. ('Sam') Patch, newly appointed AOC and, in Shrosbree's words, 'was marched into Patch's office by a group captain. "Left, right, left, right.

Cap off."' After 'the most uncomfortable five minutes of my life and a right bollocking,' Shrosbree added, Patch said, 'My punishment is that you be severely reprimanded. Do you accept this?' 'Yes sir,' Shrosbree replied. Sam Patch then said 'dismissed'. As Shrosbree was marched out, shaken by the tongue-lashing (left, right, left, right, etc.) Patch bellowed, 'Right, I'll see Shrosbree now.' The group captain grinned and handed Shrosbree his cap just as Patch called out, 'Sybil … bring us some tea. Please sit down old chap'.

The previous Thursday, Patch had been an honoured dinner guest of 601 in Town Headquarters to celebrate his appointment, and presumably Shrosbree's future had been settled there. They chatted amicably over cups of tea and Patch said that he had 'had a very enjoyable evening. You chaps certainly know how to look after yourselves'. Shrosbree apologised for putting everyone to so much trouble. The AOC just grinned and said, 'Silly bugger!'

The horseplay continued, and sometimes the public were allowed to be spectators. They would gather at weekends along the Epping Road near the south end of North Weald's main runway to watch aircraft taking off or, more spectacularly, approaching low in formation and breaking and circling to land in stream. Occasionally, after landing his Meteor and rolling onto the perimeter path at the end of the runway that ran parallel to the road, a pilot would delight the crowd by unfastening his straps, standing in the cockpit, and turning sideways to give a smart salute, allowing his plane to trickle along at idle power.

Less benign was an occasional act in which a man was seen to fall out of an open-cockpit Tiger Moth. For this a pilot, clearly visible in the front seat and waving to the crowd, would take off with a life-size dummy pushed out of sight in the rear seat but not strapped in. The pilot would perform normal aerobatics and then, while inverted at the top of a loop, push the stick slightly forward to topple the dummy out. It was a dirty trick. Everyone knew that.

Chapter 13

The Lost Weekenders

And you already know
Yes you already know
How this will end

Song by Devotchka

The argument that part-timers couldn't be expected to handle safely or efficiently the latest operational aircraft was as old as the Auxiliary Air Force. They could fly bombers but not fighters, until the Demon came along; single engines but not twins, until the Blenheim; biplanes but not monoplanes, wooden machines but not metal, slow but not fast; they were useful as a 'pool' behind the fighting line but not as a combat unit per se. Until the war. Then the Auxiliaries somehow shot down the first enemy aircraft, sank the first submarine and accounted for over a third of all enemy aircraft losses during the Battle of Britain. Any official hint that the Auxiliaries were superfluous or should be confined to a minor rôle would have been unthinkable. In November 1954 the unthinkable was said, and it was the turn of Grosvenor's followers to be surprised.

The new bogey was the swept wing, supersonic fighter. At first it was thought the familiar skeleton was being taken from its cupboard whose bones were heard to rattle whenever a new type of aircraft came into service. But soon it became clear that this was no skeleton and that harsh economics lurked behind the new official attitude to non-regular forces. The nation was hard pressed to maintain its regular forces, the cost of new aircraft was great, and it would be cheaper to reequip the fighter defence if there were fewer pilots. Just as economics had given birth to the Auxiliaries, now it was to be their death knell. Harold Macmillan, then Defence Minister (and tagged 'Mac the Knife' for the sweeping defence cuts he proposed), announced in a speech to the House that it wouldn't be 'possible or right' for Auxiliaries to switch to the expensive new machines, a necessity if they were to remain in the front line of defence. He added that 'the government has decided to alter the organisation of this force to enable those Auxiliary pilots who can give their time to it to train on the swept-wing aircraft themselves as individuals; not to equip the squadrons with these machines, but to train the men. By this means they will provide reserves behind the regular squadrons in war.' They would retain their town headquarters and their present airfields with a purely training flight of Vampires and Meteors.

The intention, in short, was to abolish the RAuxAF as such and attach its members to the regular force until the supply of national service pilots dried up and the last resistance collapsed with the dissolution of squadron entity. Certainly the squadrons couldn't survive such a transplanting for their roots were in their histories, not in the hangars and crew rooms of any aerodrome. To abolish pride of unit was to abolish voluntary service. This was well known, and the scheme met with outcry in the Commons, the Lords and the press. Informed aviation correspondents interpreted the move as a feeler towards the formal abolition of the Auxiliaries, but the tactics were understandable; any government wishing to dismantle the force at such a time would have to proceed gingerly.

After Macmillan had admitted in debate that his proposals implied the withdrawal of twenty squadrons from the front line order of battle, parliamentary opinion hardened, and modifications to the plan were introduced. The squadrons would be permitted to keep their full strength of Vampires and Meteors for the subordinate but nonetheless operational rôle of low-level defence, but they wouldn't be reequipped with Hunters or Swifts.

The shock with which their plans had been met surprised even the planners. No doubt they imagined that by allowing the force to retain their town headquarters and territorial connections the change would be presented in digestible form. The bait fell lightly enough, but the swirling current of political opinion tore the rod away. Next time the fish would have to be hit on the head.

It was two years before the government struck again, two years during which for all their high spirits and intensive training the Auxiliaries' morale was unable completely to recover. The regular squadron at North Weald, No. 111, exchanged its Meteors for Hawker Hunters, and the Auxiliary squadrons, 601 and 604, did not. Worse still, worn-out Meteors that were sent to the maintenance units were replaced with earlier, inferior versions. It was obvious that this couldn't go on for long.

In October 1955 the Legion was moderately encouraged by the appointment of Sir Dermot Boyle, who had taught Sassoon to fly, as CAS.

'In addition to its many other achievements,' he telegrammed 601, 'Legion can now take credit for having trained CAS.'

But the undertow of rumour strengthened. The squadron felt that it was sitting on a chair from which attempts had already been made to dislodge it, and it tapped the grapevine systematically, though without enthusiasm. By the winter of 1956 it knew that drastic cuts were in sight to reduce or eliminate the Auxiliaries' burden on the taxpayer. Judging from long experience, it seemed certain that only a sudden and acute threat to Britain's national security could prevent this. In November 1956 exactly such a threat arose.

The Suez affair, however, had totally the opposite effect from all previous international crises, and instead of reversing matters it accelerated them. It strengthened, with the severe drain on Britain's gold and dollar reserves, the hand of Macmillan and the politicians who spoke of cutting out what couldn't be afforded, and it made 'Mac the Knife' prime minister. The first details of total disbandment of the Auxiliary flying units began to leak through the newspapers. Max Aitken bannered the *Sunday Express* with the news on the front page. In the *Evening News* a cartoon showed a club flyer chatting at the bar and saying to another, 'That's modern warfare for you – some bright Treasury type presses a button on his desk and bang goes the Auxiliary Air Force!'

As no official comment could be wrung from the authorities, discussion in the press was protracted and joined from almost every angle. Several letters to the editors voiced a relief that the 'petrol-wasting joy riders' would soon leave them in weekend peace; some writers seemed to believe their homes had been targeted: 'They have made mincemeat of my nerves with their power dives over my chimney pots, and I've grown so tired of ducking every few minutes while trying to dig my garden.' Ducking indeed?

A flood of letters in the newspapers took up the case of the part timers. 'My heart bleeds for Mr H—,' somebody wrote. 'He might feel better if he were to go up to the Abbey, look through the book of Remembrance, and take note of the number of Auxiliary Air Force men who gave their lives for us. They had all been "joy riding young men" at weekends.'

Things began to happen quickly. On the evening of 9 January Sir Anthony Eden resigned the premiership. The following morning Clive Axford, one of 601's national service pilots, when preparing to take off from Renfrew in Scotland for North Weald after a navigational flight, was shown an Admiralty letter to the naval reservists at Renfrew banning all further flying with immediate effect. When he reached North Weald the RAuxAF equivalent of this message had not yet been received, so he made another flight after lunch to become the last Auxiliary to fly. The message came through while he was airborne, at 3.30 p.m.

The COs were summoned to a meeting at Air Ministry the next day, 601 represented by its latest and last commanding officer, Squadron Leader Peter Edelston, AFC, a slightly built advertising executive who had recently succeeded McCarthy-Jones. He was considered the best formation leader in the squadron. In the chair, Sir Dermot Boyle was flanked by the Commanders-in-Chief of Fighter and Reserve Commands and three civil servants.

The CAS came straight to the point. The squadrons and any of the ground units (radar control and interception) would be disbanded on 10 March. All flying had stopped and wouldn't be resumed, except for essential air tests, which were to be carried out by the regular members of the squadrons. It would have been bad enough for the commanding officers to have to return with such news to their squadrons, but they couldn't even be allowed to do this. The prime minister's resignation had precipitated a cabinet re-shuffle, and not until a new air minister was appointed to replace the outgoing Mr Nigel Birch, and had made an announcement to the House, could the squadron members be informed.

It was ten days before a new air minister was appointed, and the choice was ironic. George Ward, who had contributed so much to the Legion's pre-war life and been a frequent visitor to its dinners since the war, as the new minister had now to preside over its dissolution as the first job at the highest point of his career. He showed no relish for this responsibility. At a dinner he removed his jacket and rolled up his shirt sleeve to show the Flying Sword tattoo. 'That's the way I used to feel about the Legion,' he said simply, 'and it still is.'

The squadron's constant and sometimes almost embarrassing ally, Max Aitkens's *Daily Express*, attacked the CAS. Under the heading, 'A Tale of Two Bs' in its Opinion Column, it fulminated:

> This is the sorry tale of two Bs – Air Minister Nigel Birch and his Chief of Air Staff Sir Dermot Boyle, the men who are now scrapping the RAF's Auxiliary squadrons. For Mr Birch there is some excuse: he had never known much about the RAF anyway. But for Air Marshal Boyle there is no excuse at all. He was an adjutant of an Auxiliary squadron himself … These proud wings on his tunic should be drooping low in shame.

Boyle made it clear that nobody believed the Auxiliaries incapable of flying the latest fighters with perfect competence. Severe cuts were in store for the regular force also. A press handout issued by the Air Ministry contained one sentence in its four hundred words that encapsulated its attitude to 'low level defence' (under thirty thousand feet): 'Technical development is continually reinforcing the view that the main threat to this country is from high altitude attack; other forms of attack which might be encountered by less advanced types of aircraft are in consequence less likely to develop.'

In the Royal Air Force Club in Piccadilly the Auxiliary commanders formed a committee of the four metropolitan COs to fight the decision on behalf of all twenty squadrons. The aim of this unusual alliance was to influence public opinion. This hope was indeed dim, but there were two months to go with no flying, and these were not passive men. The first move was a television appearance at the Pathfinder's Club in Knightsbridge, when they answered questions put by the BBC air correspondent. The second was a letter circulated to all members of parliament and all national and provincial newspapers, explaining the role of the Auxiliaries.

Political leaders were approached – Air Commodore Vere Harvey, head of the Government Air Committee, and Geoffrey de Freitas, Labour ex-Under Secretary of State for Air. To questions in the House, however, George Ward reaffirmed the government's unswerving intention to carry out its unwelcome duty in the interests of national economy.

The Auxiliary committee was received everywhere it turned with goodwill and sympathy, but it was beaten. Not even a 'Ned' Grosvenor could help. Now that the die was cast and there remained no basis for negotiation, all doors were open, official sympathy was boundless. 'How are the four politicians today?' Dermot Boyle would enquire if he met one of the Auxiliary committee members.

The dawn of 6 March 1957 was clear and inviting to the aviator. It was the day of the last parade. Kensington Park Road was closed to traffic as the officers and men marched out from Town Headquarters at 4.15 in the afternoon and halted at the markers. Quietly, HRH Prince Philip's limousine pulled into the curb and an order brought the squadron to attention. The Flying Sword flag with a cross of St George designed by Grosvenor stirred at half mast as the Prince inspected the parade and mounted the saluting base to address the hundred men before him.

'Before you march past for the last time,' he said, 'I just want to say goodbye; and, as I expect no one else will say it, to thank you for the service you have given to this squadron, to the Auxiliary Air Force, and also to your country in peace and war.'

'As I expect no one else will say it!' Few people would understand how well received those few words were.

At a cocktail party after the parade the Prince was shown, in a darkened lecture room in Town Headquarters, a mock catafalque made by the airmen on which reposed in state the life-sized effigy of a pilot in flying kit, his gloved hands symbolically clasping a spent champagne bottle. Palely illuminated by candles stuck in the necks of 'Vat 69' bottles, a notice reflected the other ranks' bitter disappointment:

HERE LIES THE BODY OF 601 SQUADRON
ROYAL AUXILIARY AIR FORCE
MURDERED BY IGNORANCE

Wreaths and messages surrounded the body: 'Peace, all is Peace – from the Epping and Ongar District Council'; 'In Deepest Alcoholic Remorse – from the Epping and Ongar Licenced Victuallers'; 'With sincere gratitude to the Air Ministry – from Bulganin and Kruschev, the Kremlin'; 'In Memory of happy nights – from The Young Ladies, the Egyptian Queen, Malta GC'.

It was 10.30 by Dickie Shaw's clock under Grosvenor's portrait when the last officer slammed the door behind him. Under the porch and through the arched doorway passed for the last time city executives, civil servants, salesmen, farmers, students, policemen, and bus drivers whose other lives had been spent flying or servicing aircraft, and who had now to find another way of joining Friday to Monday. They would succeed, but it would take a little time and a little sorrow. Punch displayed a full-page drawing by Brookbank of three pilots shuffling disconsolately away towards the sunset, their Meteors parked behind them, glancing up at a swept-wing fighter streaking across the sky. The sentimental caption read:

When we could serve by flying we gave up our time and flew;
But now, it appears, our country has nothing for us to do.
No doubt the boffins have got it taped, but we'd like to make it clear –
If they every find a need for us, they'll always find us here.

With the hindsight of half a century it is possible to see that in the long run the government was right, and that Britain's long devotion to amateur service had run its course and would have to yield to the reality of technology.

Flying ability was no less important but flair and courageous individualism would yield to technical mastery and relentless learning and practice. Brian Thynne's repugnance at the replacement of horses by mechanical vehicles was to be re-lived in the twenty-first century with the shock of pilotless drones replacing human pilots. Also, the capital cost of aircraft and flight equipment from parachutes to pressure suits would hardly pay off with only partial use.

No. 54 Kensington Park Road was for some years obscured from the road by trees, a clean patch over the archway marking the removal of the Flying Sword. This, Brian Thynne's gift to the squadron, was sent to No. 24 (City of Adelaide) Squadron, an Australian Auxiliary unit with which 601 had affiliated, at the request of the governor general. The entire block has since been demolished and replaced with high-end flats.

There were withdrawal symptoms for many squadron members at weekends. A few joined the RAF's air control units but found controlling an aircraft by staring at a green screen no substitute for controlling one from the cockpit.

An exception was the Hon. Peter Vanneck. The son of Lord Huntingfield, to say that he was an interesting character would be to observe that zebras have stripes. Enigmatic and opaque, he combined gravitas with eccentricity and a perpetual, self-deprecating sense of irony. Though he had been raised in Heveningham Hall, one of England's most beautiful houses and perhaps as grand and splendid as Sassoon's Trent Park, and where according to Vanneck 'the kitchen was a couple of cricket pitches away from the dining room', being the second son he now lived in a four-bedroom bungalow, White Lodge, in a small Suffolk village. He wore naval wings on his highly individualised RAF uniform, while in the mess a gold swizzle stick for stirring cocktails hung from his collar. He drove a dilapidated pre-war Austin taxi and was fined ten shillings for leaving it parked on a slope outside Epping tube station, prompting suppressed laughter in court when he explained that he did this in order to be able to start it from cold when he returned late at night. It was by choice that he had risked his neck landing planes on carrier decks with the Fleet Air Arm and was now doing so flying jets at weekends and at other times racing his private red-and-black, Flying-Sword-emblazoned Tiger Moth, G-ANOD. He once emerged from a Vampire at Horsham St Faith, Norfolk, for a shoot, replete with tweeds and gun.

Where others saw boredom in a controller's job Vanneck saw opportunity. After transferring to No. 3619 (County of Suffolk) Fighter Control Unit he climbed its ranks to Air Commodore and succeeded in meshing his service and civilian lives with personal charm to become deputy chairman of the London Stock Exchange, Alderman of the City of London, then Sheriff and finally Lord Mayor of London. At the pinnacle of fame he remained typically solemn and silly at the same time. Describing his normal day as Lord Mayor for a magazine he was a caricature of himself: 'I take my first pinch of snuff with the first problem which is probably about now. It is Blend X, Judge Carmel's mixture. Snuff, like China tea, which Cordelia and I love, should be mixed. I use a silver snuff box by day and change to gold and Royal George a night …' Air Commodore Sir Peter Beckford Rutgers Vanneck, GBE, CB, AFC died in 1999.

Denis Shrosbree, his record clean despite the Twickenham incident, became a BOAC and British Airlines captain and, after reaching that airline's retirement age, a captain in Virgin Airlines. His partner in risky aerobatics, Desmond Norman, ploughed ahead with his company but faced cost problems with developing a three-engined Trislander and the government called in its development loan. Britten-Norman Ltd passed into other hands and continued to operate. Desmond Norman CBE died in 2002.

The youngest Norman brother, Torquil, did his national service in the Fleet Air Arm, narrowly escaping from a fiery crash in a Hawker Sea Fury before joining his brothers on 601. Just before the graduation ceremony at Syerston where Torquil received his wings, a Meteor 7 landed and his brothers Mark and Desmond stepped out, attired in business suits, bowler hats, and conspicuously red socks. After disbandment he worked as an investment banker in New York, crossed the Atlantic multiple times in both directions, usually solo, in an assortment of light aircraft: two twin-engined Piper Comanches, a twin-engined biplane DH Dragonfly, and a Leopard Moth. He flew over three thousand miles non-stop from Bermuda to Lisbon in a Comanche, which took twenty-one hours, with his soon-to-be wife Anne. After creating and then selling a highly successful toy company, Torquil donated a large portion of his wealth to converting 'the Roundhouse', a decaying theatre in Camden, into an arts centre for under-privileged young people. For this he was knighted, which with older brother Sir Mark made two in the family.

In June 1967 the *Sunday Times* reported that Dr Nevil Leyton, the migraine and headache specialist, had been found guilty by the General Medical Council of 'infamous behaviour in a professional respect' by self-advertising and soliciting investment in a non-existent drug company. His refreshing response was, 'Why wasn't I struck off the Medical Register? That's what I wonder myself.'

In 1977 Sir Max Aitken was forced to sell his birthright, the once mighty Beaverbrook Press, which had fallen on bad times. The buyer was The Trafalgar House group, and the price was rock bottom. As part of the deal Aitken was provided with a modest Ford Escort – his only perk. Like its predecessor, a Rolls-Royce Corniche, the car's licence plate read 'RAF 601'. After suffering several strokes Aitken hired a chauffeur. But there was a stipulation that he would return the car when it reached a specified mileage. When it did, Trafalgar's principal, Victor Matthews, who had risen in the newspaper business from a sixteen-year-old office boy and served as a seaman in the Second World War, showed the single- if not in his case the narrow-mindedness of a self-made man and demanded the return of the car, unmoved by Aitken's pleas that it was now his sole means of getting around and that he paid for his own petrol and chauffeur. It was the only battle Aitken ever lost, and he had to forfeit his high-mileage Escort. Sir Max Aitken, former multi-millionaire and champion sportsman, yachtsman, war hero, credited with shooting down sixteen enemy aircraft, winner of the DSO and DFC, MP, renouncer of a Lordship, post-war negotiator between the British government and Rhodesia thanks to his acquaintance with the Rhodesian prime minister and Spitfire pilot Ian Smith, died in 1985 aged seventy-five.

Chapter 14

Sequel

You'd be surprised if you could hear
How very loud a ghost can cheer –
'Good-bye then now, and thank you boys!'

Flying Officer Barnes, 601 Squadron
Intelligence Officer, 1940

Allied soldiers in the Second World War drawn from civilian life were loath to shoot and kill and only a minority did so. But aerial warfare masks brutality. Every fighter pilot is a killer but is spared from seeing his victim. Billy Clyde told his wife, Barbro, that 'firing at an aircraft is different from shooting a man'. No. 601 pilots resisted AVM Dowding's order to shoot at Luftwaffe aircrew being rescued by German seaplanes bearing red crosses. (The order complied with the Geneva Convention, since pilots being recovered to fight again were still combatants.) Australian Bill Matheson thought little of blasting a passenger train in Italy, but when chasing an Fw 190 that was twisting and turning was startled suddenly to see the pilot in the cockpit. Shooting down an aircraft doesn't always mean killing its occupants, but it usually does. Huseph Riddle, tempered by the sensitivity or more likely the imagination of an artist, openly confessed that he deplored the killing part of his job.

It took intense concentration and disregard for personal safety as well as cold blood to shoot accurately. High-scoring fighter pilots on both sides learned to sneak up close behind an unsuspecting enemy and open fire at point blank range from which there could be no escape, an act that has been termed, 'murder in the air'. It is hard to think of the cultivated and decent men of No. 601 Squadron in the peaceful years before the First World War, gentlemen in the best sense, as killers, but that, to Germany's surprise and Britain's good fortune, is what they were.

After baling out of his Hurricane blinded from Perspex fragments in August 1940, 'Mouse' Cleaver underwent eighteen operations to remove the fragments. Only in one eye was sight preserved, but the operations had an unexpected consequence. The discovery that the human eye didn't reject Perspex led the surgeon, Sir Nicholas Ridley, to develop artificial lens replacement surgery for cataract patients, to the eventual benefit of millions of people all around the world (including this author). When the vision in his good eye began to suffer from a cataract, Cleaver himself underwent the very operation his Battle of Britain wound had made possible.

Billy Clyde was shot down in the same air battle as Cleaver. Back in England, after service in Washington with the joint chiefs of staff he discarded his wing commander's uniform and sailed for New York to resume work as an executive with Johnson and Johnson. He travelled extensively and contracted a severe attack of colitis, thought to have stemmed at least in part – he believed entirely – from his long exposure in the English Channel, and the doctors said it would kill him in six months if he didn't take a long break. He planned on retiring to Sri Lanka for his health but friends visiting him on their way to Acapulco on Mexico's Pacific coast persuaded him to go there with them instead. He did so for a six-month convalescence but after golfing, swimming and sailing found he liked it so much he never left. Ever an over-achiever, he even became Mexico's national champion in dinghy racing.

Clyde fell in love not only with Mexico and his new lifestyle but with Barbro, a half-Jewish Swedish woman twelve years younger who had escaped with her family from Prague in1939 as the Germans marched into Czechoslovakia. Clyde was by now divorced, soon Barbro was too, and they got married.

But because he still had to earn a living he asked himself over and over what delights the world had to offer that Acapulco lacked. One day the blindingly obvious answer came to him: *duck a l'orange*! So his first venture was a duck farm. With two thousand birds in the middle of a residential development, he sold birds to restaurants that were then able for the first time to offer the item on their menus. But ducks are noisy, and two thousand of them were too much for the neighbours, so he had to close the duck farm. Then he found the perfect channel for his entrepreneurial and social skills, or rather it came to him.

The town, situated on a wide, semi-circular bay surrounded by tall hills, blessed with white beaches and a sunny climate and closer to Los Angeles than is Florida or New York on America's East Coast, was proving in the 1950s to be a magnet for Hollywood stars and film industry magnates. Frank Sinatra was singing, 'Come fly with me … down to Acapulco Bay'.

These new visitors found it difficult to arrange housing and car rentals, engage tradesman, and obtain their favourite food and liquor. Clyde was frequently contacted for help with these as well as for advice on finding a villa or apartment to purchase or rent. He began to charge a fee for such services and graduated to residential construction and renting. Though he wasn't an architect he had supervised the building of his own house and this gave him sufficient confidence. Gradually he grew a business out of developing properties for resorts or luxury houses. He promoted Acapulco energetically, in his element, readily assimilating with the incoming celebrities.

Clyde seemed to know everyone, and man of the world would seem an apt description. Among the celebrities drawn into Clyde's circle was his pre-war friend David Niven. After resuming his acting career in 1945, Niven was voted the second most popular British film actor. The following year his wife, Primula, whom he had met at The Ship, died.

Barbro recalled:

> David Niven was a very close friend of Billy's, but he also liked pretty girls and was quite a womaniser. He had many affairs: with Rosie, Billy's first wife, and Eliza, Billy's sister-in-law., etc., so Billy sent him a telegraph saying, 'This is my mother's address' and David didn't take it very well, so later when Billy and I were in NY we had lunch with David and his very pretty Swedish wife [divorced fashion model Hjördis Paulina Tersmeden]. David sat next to me, but he barely spoke to me for fear of upsetting Billy. It was just a big joke that David took seriously.

After Billy died in 1984 Barbro moved from Acapulco to Mexico City. One day she received a significant visit.

The middle-aged man to whom Barbro opened her door had been put in touch with her by Ed McManus in England, then a member of the committee for the 2005 Battle of Britain Monument. While acquiring names to be engraved on the monument McManus noticed that two addresses in Mexico City related to 601 Squadron. Barbro's apartment 501 on the fifth floor in Polanco, an upscale area of the city, was large and bright. The first thing that struck Paul was how stunning Barbro was, even in her eighties. Slim, of medium build, he had no doubt that she had been 'a corker', as he put it, all her life, and no doubt a match for Billy the playboy. Her manner was charming, her speech accentless and seductive. Then Paul noticed a framed photograph on the wall, of Billy Clyde with another RAF officer, an unusually tall man even allowing for the contrast with 'Little Billy's' diminutive stature. Both were smiling; they had just shared a Heinkel. On looking more closely Paul recognised the other officer, although he had only seen him in photographs. Sure enough, it was his father, Michael Doulton.

When Flying Officer Michael Doulton was shot down on the 31 August 1940 somewhere over the Thames Estuary, his wife, Carol's, letter to her father in America explained the plan she had solemnly agreed to with Michael in the event that he was killed, as they both anticipated, which was that she would have her baby in America, 'safe from war and bombs'. The United States was still neutral as was Portugal and she made her way with a flight to Lisbon ('full of Germans and spies') and sailed on the American Export Line to New York, landing there on 12 October. Carol's son, Paul, was born on 14 April 1941, not quite on April Fool's Day as she had whimsically predicted. In January 1945, keeping her pact with Michael that their child would be raised 'as an Englishman', Carol returned with Paul, now four, in a freight convoy escorted by naval vessels from Halifax, Nova Scotia, the risk from submarine attacks now being low, arriving at Liverpool three weeks later. Settled in Sussex, Carol taught at a boy's prep school and became curator of the Battle Museum, which commemorates, ironically, the last successful invasion of England, by William the Conqueror in 1066.

The location of Michael Doulton's crash site was uncertain. Two Hurricanes from the same flight of four had been seen falling at the same time onto either side of the Thames Estuary. One was known to be buried deep underground in a location near Wennington Church, south of Romford, Essex, but because many crash sites had been disturbed for souvenirs or scrap metal (even today you can buy Hurricane 'relics' on the web) the Ministry of Defence banned further excavations. A Czech pilot, Jaroslav Sterbacek, had been shot down in the same dogfight and within five minutes of Doulton, and a third pilot claimed to have seen Doulton's aircraft splash down in the Thames, so the Wennington wreck was assumed to be Sterbacek's. With this identification the Ministry of Defence gave historians permission to excavate, subject to the family's approval. However, contact with Sterbacek's family in Czechoslovakia revealed that they had the wrong pilot; possibly therefore it was Doulton. On 27April 1984, at the expense of Carol's father and with renewed permission of the Ministry of Defence the aircraft was raised and identified as Michael Doulton's Hurricane R4215. His remains were still in the cockpit, the gun button on the control column still set to Fire and the throttle fully open. He was cremated in a private ceremony at Hastings and his ashes interred in Salehurst churchyard, East

Sussex. This was the first and only time Paul ever spent in the presence of his father. Carol died in 2006. Paul Doulton had lived in Mexico City since 1980, and was still living there in 2014, managing Wellcome Company's pharmaceutical company's operations.

Many of those lost were less well remembered. Four 601 pilots died in the same dogfight on 11 August off the Dorset coast. The body of newly qualified twenty-four-year-old Flying Officer Smithers drifted to the French coast and he was buried at Le Havre. In 2013, over seventy years later, John Wheeler received via his 601 website a photograph of a uniformed RAF officer from Cheryl Ludgate, a lady in Dorset. On its back was written 'Julian Smithers', and a web search had led her to the website. Ms Ludgate had found the photo at a jumble sale, where it had been recovered from a garbage skip.

Loel Guinness was posted as operations controller at No. 13 Group when controllers with fighter experience were sorely in demand. After D-Day he commanded No. 145 Fighter Wing of the Second Tactical Air Force in France, being mentioned in dispatches five times. The Netherlands made him a commander of the Order of Orange Nassau and France awarded him the *Legion d'honneur* and the *Croix de Guerre*. He had the rare satisfaction of liberating his own chateau from the Germans during the Allied advance. His father left him a fortune on his death after the war, with the admonishment to go abroad and keep moving to avoid oppressive taxes. This Loel did, patriot though he was. He set up a home in Vaud, Switzerland. In 1951 he and Lady Isabel got divorced and he married Gloria Rubio, a Mexican who had been the wife of Prince Ahmed Fakry of Egypt. Drawn into Billy Clyde's glittering orbit they moved into an estate in Acapulco, where Loel Guinness died in 1989 aged eighty-two.

Aidan Crawley survived PoW camp, married Virginia Cowles, wrote numerous books including *Escape from Germany 1939–1945*, founded Independent Television News (ITN) in 1955, and became one of the first BBC television documentary-makers on current affairs. He didn't rejoin the Legion but plunged into writing, broadcasting, and politics, and in what appeared to be a 601 tradition since Sassoon was appointed Under Secretary of State for Air in the post-war Atlee government. Then his life descended into tragedy. Virginia died in a car crash in France in 1983. (Nigel Nicolson recalled in his memorial address that the first time he met her, 'she was the most beautiful young woman on whom, until then, I had ever set my eyes'.) Five years later both his sons died in a plane disaster. He lost heavily in a Lloyd's crash. Aidan Merivale Crawley, sportsman, fighter pilot, prisoner of war and escaper, politician, author, journalist, television executive and broadcaster, died in 1993.

Archie Hope, managing director of Alvis Motor Cars when interviewed in the 1950s, was a polite and modest man, hard to visualise as the tough and seasoned fighter leader he had been. He still nursed frustration over having lost time because of the order, eventually disobeyed, not to attack the Stukas bombing Tangmere. Group Captain Sir Philip Archibald Hope, CBE, DFC and two Bars died in 1987.

Squadron Leader E. J. Gracie DFC, 601's CO in late 1940 and wing commander in Malta and Italy, was killed in February 1944, aged thirty-two.

Both Jack and older brother Huseph Riddle flew with 601 throughout the Battle of Britain. Jack had scored one victory before joining 601 and one after, while Huseph Riddle scored a single kill with 601 on 11 July 1940 and more thereafter. Both Riddles were promoted to squadron leader on the same date in 1941. Huseph, for all his combat success, later confessed to having been uncomfortable with the lethal aspect of aerial warfare. He became a controller in Scotland, respected for his knowledge and air combat experience by the pilots he guided to interceptions. After the war he became a famous portrait painter and a member of the Royal Society of Portrait Painters. Among his many notable works was a portrait of Queen Elizabeth for the RAF Regiment. Later he resided permanently in France.

Jack Riddle's interest in the squadron was re-awakened by the 601 enthusiasts Ed McManus, Jamie Ivers, and John Wheeler. He attended a memorial to Billy Fiske when, on 17 September 2008 a stained glass window, paid for by the Old Comrades' Association, was dedicated to Fiske at Boxgrove Priory, near Tangmere, where he had lain buried in a neglected grave since 1940. A handful of surviving Legionnaires attended the ceremony, to which a replica of Fiske's Bentley was brought. Jack Riddle, ever faithful, was present in a wheelchair.

In a letter, Jack wrote of Fiske,

> Billy liked to talk with everyone around him, particularly the ground crews. He wanted to know them and all about their jobs, aircraft maintenance and where difficulties lay – always helpful. Very soon he managed to endear himself to the whole squadron – not just the officers, but the other ranks too. There was club a few miles away from the airfield at Tangmere and this was made our unofficial squadron headquarters. It was somewhere pleasant, overlooking the waters of Chichester Harbour, where our wives and friends could meet and

> be with each other, and wait together until we could be free from Tangmere. Billy always seemed to get there before I did – maybe his motor car was faster than mine! Usually I would find him at the club with a few friends, and there was something many of us noticed. If [a new pilot] came into the club looking slightly at a loss, possibly not knowing many members of the squadron, Billy would be on his feet. He would go over, introduce himself, find out who they were waiting for, get them a drink and suggest that they might care to wait with him and his friends until the person arrived.

After living charmed lives, wing commanders Jack and Huseph Riddle both died in 2009, Jack at ninety-five and Huseph at ninety-seven.

Sixty-five years after Canadian EATS pilot Ray Sherk escaped from France, having baled out of his Spitfire, he returned with his wife to visit his former helpers at Beaumont Hamel. He learned that Leo Roussel, who forged Sherk's papers, and Rene Munchombled, who guided him to the Pyrenees in the south France, had been executed by the Nazis. The Sherks met Roussel's wife and Munchombled's daughter. Sherk was living in Canada and still flying his own seaplane at the age of ninety. One day he received a surprising phone call. His flying club had been asked to be put in touch with him by someone in Germany. Something about a Ju 52.

The Ju 52 was the Luftwaffe's packhorse. A slow, lumbering, slab-sided, three-engined transport with fixed undercarriage, formerly a civilian aircraft of the type used by Hitler, it was vital to German transportation in the Mediterranean. Crossing from north to south, the Ju 52s carried personnel, spares, fuel, munitions, and medical supplies for Rommel's army in North Africa, returning with wounded, prisoners of war, and sundry equipment. Slow, unmanoeuvrable, minimally armed, if spotted by fighters they had little or no chance. They were prime targets for British fighters, and it was on such a crossing northwards that Clive Mayers as a prisoner en route to a PoW camp in Germany almost certainly lost his life to the guns of the RAF, possibly his own squadron. Although such aircraft were highly vulnerable to fighter attack, the Mediterranean is large and the chances of escaping notice while crossing it during the desert war were reasonable, especially given fighter pilots' well known reluctance to fly far over water.

Leutnant Horst Willborn of the Luftwaffe had flown 155 uneventful missions, including thirty Mediterranean crossings in a Dornier Do 17. In September 1942 he commanded a Ju 52 crossing to Benghazi carrying radio equipment, with instructions to fly back to Athens to report to his superiors after a stop in Tobruk en route. Willborn wasn't the pilot – Luftwaffe practice was generally to give command to the most senior officer on the aircraft. In Tobruk they took on board some severely wounded personnel for transfer to a field hospital and soldiers returning on leave, giving a total of twenty-two passengers. They flew very low to avoid detection but lacked the cloak of a vast expanse of sea and were easily spotted by Squadron Leader Matthews and his two wing men, 'Crash' Curry and Ray Sherk. The Ju 52 was attacked from the left rear by three Spitfires. The pilot put the Ju 52 into a sharp turn and the gunner returned fire but, in Willborn's words, 'It was an uneven contest.' The Ju 52 landed heavily, on fire. Everyone was pulled out, with one crew member badly injured, before it was engulfed in flames.

Since then Willborn had always been curious to know about the three Spitfire pilots and the internet made his research possible. He learned that 'Crash' Curry and Matthews had survived the war but since died, and that Ray Sherk belonged to a flying club in Ontario, Canada. He tracked Sherk down and in 2010 Sherk and his wife visited Willborn and his family in Hamburg, where they were honoured guests. Ray Sherk keeps a Piper Cub on floats in Canada moored on a nearby creek. He takes off into a bend in the river and over a bridge if the wind is strong enough, or under the bridge if it isn't.

Stanisław Skalski, 601's CO in Italy and one of the most capable and experienced fighter pilots in the RAF, was given a staff position at No. 11 Group headquarters after his tour in Italy. When the war ended he returned to Poland and joined the army's air force. In 1948 he was arrested by the communist regime on fabricated espionage charges and sentenced to death, later commuted to life imprisonment. After the end of Stalinism he was released, rehabilitated, and allowed to join the military. He served at various posts in the Polish Air Force and was promoted to the rank of brigadier general. In 1990 he had an emotional meeting with the pilot of the downed German aircraft he had landed beside to aid its wounded crew members on the first day of the war. Squadron Leader Stanisław Skalski DSO, DFC and two Bars, died in Warsaw in 2004.

Despite his shabby treatment after the Battle of Britain, Keith Park lived to see his reputation for tactics and strategy, further burnished by Malta, soar as arguably the greatest commander in the history of aerial warfare. Even his Big Wing opponent Douglas Bader admitted as much. Streets, and a locomotive, were named after him. His statue stands in London's Waterloo Place. Air Chief Marshal Sir Keith Rodney Park GCB, KBE, MC and Bar, DFC returned to New Zealand where he died in 1975, aged eighty-two.

Sir Philip Sassoon's gorgeous Porte Lympne in Kent by the English Channel fell into gradual decay after his death and was bought by John Aspinall, a zoo owner with a high respect for animals, and became the Porte Lympne Wild Animal Park.

If you walk half a block north from White's in St James's Street to Piccadilly and then ten minutes west along the north side you will come to the Royal Air Force Club at No. 128, opposite beautiful Green Park, a blue RAF flag fluttering over the entrance. The ground floor is secured with great taste, which renders the automatic doors and metal barriers barely noticeable. Membership is less restrictive by far than for White's but nevertheless limited to serving and former officers of the RAF and those with specified links to the force. Beyond the aero-masculine lobby with its long reception desk extend a virtual aviator's 'stations of the cross' revering Sidney Camm, designer of the Hawker Hurricane, Roy Mitchell, designer of the Supermarine Spitfire, and Frank Whittle, inventor of the jet engine, together with photographs and memorabilia, including flight test notes on the Spitfire II, dated March 1936, a full year before 601 rejoiced at receiving Demons. Beyond the lobby stretches an elegant, wide, and tall main corridor hung on both sides with paintings of aircraft. The club is vast, with a large dining room, a lounge and tea room overlooking the park called the Cowdray Room after the Club's founder, Lord Cowdray, a ballroom, main bar, library, billiard room, conference rooms, several meeting rooms, a computer room, a squash court, and four floors of bedrooms. Pictures in the bedrooms tastefully eschew military hardware and war. The atmosphere throughout is light and airy.

On the basement, among other pictures hangs a coloured, composite illustration about four feet wide depicting the escape from Stalag Luft III. It succeeds in including all the elements without clutter: RAF aircraft being shot down; airmen captured; the camp exterior with its watch towers and searchlights and pine trees beyond; the interior of a hut with a lookout at the window. A cutaway of the camp depicting 'Harry's' enormous scale and sophistication makes an impact that words cannot; to see its great depth and length put into scale together with the workshop at the bottom of the shaft, the half-way houses, and all the supporting devices is breath taking. Off centre of the composition is depicted the re-capture and imminent shooting in the back of two escapers. In the centre a girl searches for a name on a cenotaph. The border contains full-face photographs with names and nationalities of the fifty murdered, twenty-two from Britain, Bushell's one of them, the others Commonwealth and various European. The superb illustration, by Halder, is executed with precision and taste, devoid of excess and sentimentality.

Near to this is another relief plaque that deserves to be noticed, donated in memory of Canadian Flying Officer Albert Day. A man and a woman in civilian clothes are supporting a dazed airman who is wearing his Mae West, helmet, oxygen mask, and flying boots. In the night sky are aircraft, searchlights, and parachutes. The plaque is dedicated to 'the brave men and women of enemy occupied countries' who at great personal risk helped 2,803 downed Allied airmen. While the importance of the Resistance in the overall war was negligible, at the individual level it was immense. It smuggled many airmen home, running grave risks that British civilians never faced. Whitney Straight, Roger Bushell, and Ray Sherk from just one squadron, and possibly others if they had lived to talk, would have attested to that.

Next to this area is a business centre, and opposite a pub called The Running Horse, serving drinks and meals, bustling at lunchtimes and evenings, and here a declining handful of ex-601 pilots, all post war, gathers informally every year or so, sometimes with their wives. One floor above the main corridor is another one displaying rows of squadron crests, hundreds of them, on both walls, round a corner and down the wide back staircase. No. 601 (County of London) Squadron's crest occupies a nondescript location opposite the lift, one of so many, nothing remarkable. Except that to the observant one thing does stand out. Always different, 601's is the only crest without a ribbon or motto.

On 13 May 2009 a ceremony was held at Hendon, the self-proclaimed birthplace of British aviation, to commemorate No. 601 Squadron as the airfield's longest serving RAF unit and the place of its origins. It was attended by AVM Ian McFadyen and the deputy mayor of London together with former officers and members. Billy Fiske's magnificent green Bentley was present, not a replica this time, procured on loan by Jamie Ivers.

This wasn't easy to do. When Jamie tried to track the car down through the Bentley Owners' Group and explained that he wished to borrow it and why, he drew a blank, except for a vintage car dealer in London who responded to ask about Fiske for more history on the car because its owner was offering it for sale at £2.5 million and a celebrity pedigree would boost is value. Jamie provided information on Fiske and asked if in return he could borrow the car for the worthy purpose of the dedication. The dealer promised to pass on the request. Surprisingly, the Bentley's owner agreed to lend it for nothing and even to pay the transport costs but insisted on remaining anonymous. It isn't known who that owner was.

After the unveiling and speeches tea was served next to a table bearing the squadron's freshly polished silver, which had been donated by Sassoon and some porcelain dishes bearing the Doulton company's trade mark. Today, visitors to the RAF Museum in Hendon can see to the right of the main entrance a sculpted Flying Sword standing on a curved plinth a few steps to the right of the main entrance with a metal plaque in front of it. The shiny plaque is so blindingly reflective it requires an enhanced photograph to be readable, but that may be as well because its words are sadly, indeed shamefully, remiss:

> Number 601 (County of London) Squadron
> Auxiliary Air Force 1925–1947
> Royal Auxiliary Air Force 1947–1957
>
> This sculpture commemorates No. 601 Squadron's connection with RAF Hendon where it was stationed intermittently between 1927 and 1949, longer than any other RAF unit.
>
> Founded in the era of biplanes as a day bomber unit in 1925, it gained renown during the 1930s, becoming a fighter squadron just before the First World War. Flying Hurricanes during the Battle of France and the Battle of Britain in 1940, it was posted to North Africa with Spitfires in fighter and ground attack roles in 1942 and ended the war in Northern Italy.
>
> Reformed on Spitfires in 1946, it converted to jet aircraft, Vampires in 1949 and from 1952 flew Meteors until disbanded in 1957.

It is a fitting memento – almost. Not that anyone is likely to notice. There is no mention of Malta.

Chapter 15

Afterword

Fifty Pounds – A Frightening Sum!

After the war Auxiliaries were paid by the day served at the regular RAF flying rate for their rank plus travel expense and an annual bonus of £35 if performance criteria were met. This alone would be no motivation for extending eighteen months of national service by three more months of jet conversion and committing every weekend and two weeks a year to training.

Still, the income was valuable. Just when I'd decided that selling telephones wasn't for me I won a huge commission of £450 (equivalent to £10,000 in 2014). Calculating that with this and my Auxiliary earnings I could just last the four years it would take to get a degree, I took the plunge and enrolled in the London School of Economics. At the age of twenty-four this was a tough call. Then disaster struck. The engine of my 1938 Sunbeam Talbot, which at its best had the power of a lawn mower, gave in completely. The garage's quote for reconditioning the engine was £50, a frightening sum.

That Saturday at North Weald when I arrived early Mark Norman was already there, looking, he said, for four pilots to deliver Vampires to the Burmese air force in Rangoon for his company, Airwork Limited, the company his father had founded in the 1930s. It would pay £50. (I realise this would sound more credible if the timing and amounts were different, but this is how God arranged it.) They were Vampire T-11 trainers with side-by-side seating, factory-new and with freshly painted Burmese air force markings and long-range tanks. We strapped our modest luggage in the right-hand seats and took off from Hatfield early one morning in March 1955, with Chris McCarthy-Jones, our CO, leading and me as his number two, followed by George Farley the adjutant and Jimmy Evans. The routes to Istres in the South of France and thence to Malta were familiar, and from Malta to El Adam in North Africa, Nicosia in Cyprus, and Habbaniyah in Iraq gave us no trouble.

It was on the next leg that we began to earn our money. Night falls faster at this latitude than we were used to and it was already dark on the ground before we were half way to Bahrain. A line of cumulo-nimbus lay across our path and their tops, still glowing in the sunset, were far higher than our altitude of thirty thousand feet, which meant trouble. As we plunged into the turbulent middle of these towering columns, with lightning turning them on and off like giant Christmas lanterns, we were smacked and buffeted with such violence it was difficult to hold onto the stick. Outside the cockpit the clouds were alternately blinding and black, flashing from within and without. Now and again the monstrous columns of cloud parted to reveal ghostly canyons, flickering with light, that would have been wondrous to observe under different circumstances. Hail striking the plane sounded like machine gun bullets. Chris's aircraft was rocking furiously from side to side, as was mine; I don't know how I managed to keep station and probably shouldn't have tried. We landed in pouring rain on a flooded runway, my right hand locked on the stick, almost paralysed.

For five days we kicked our heels in Bahrain under almost constant rain, waiting for the inter-tropical front to pass through. When it did we were still stuck because the runway at the next refuelling stop at Sharjah near Dubai was under water and closed. But with the passing of the anti-cyclone the wind began blowing eastward strongly and measured up to 70 mph at thirty thousand feet. If that tailwind held we could overfly Sharjah and just make the thousand-plus miles direct to Karachi, Pakistan.

Bahrain is a coral island and the sea can rise up through the porous ground at high tide. After a ragged take-off in clouds of spray from water up to our wheel hubs we just made it to Karachi. From there to New Delhi to Calcutta to Rangoon was uneventful. The Burmese were pleased to see their aircraft but shocked at their condition: hail had stripped paint off all the leading edges and removed the Burmese air force markings. We left for home by BOAC that night, reaching Heathrow the next day. I took a train to Blackbushe airport, handed over my papers, and received £50 in an envelope. Another train, two tubes, two buses, a moderate walk, and the money was gone, but I had my little Sunbeam Talbot back and functioning.

So I got my degree in Economics, wrote this book's predecessor, *The Flying Sword*, and went to work for two giant multinationals, one Swiss (Nestle) and the other American (Ford). For the next forty years I worked with many Germans who had been children in cities devastated by the RAF, including my marvelous secretary Helga Schultze from Cologne. Some had served in the Wehrmacht, like Helmuth Heilner, a Jew, who said this was the safest place for a Jew to be.

And here was the strangest thing: I found them all to be wonderful people.

Bibliography

Barnham, Denis, *One Man's Window*, W. Kimber, 1956.
Bungay, Stephen, *The Most Dangerous Enemy*, Aurum, 2010.
Channon, Sir Henry 'Chips', *The Diaries of Sir Henry Channon*, Orion Publishing Group, Limited, 2000.
Crawley, Aidan, *Escape from Germany, 1939–1945*, Collins, London, 1956.
Dean, Sir Maurice, *The Royal Air Force and Two World Wars*, Cassell, 1979.
Deighton, Len, *Fighter*, Book Sales, Incorporated, 2000.
Flypast Magazine, 'Desert Encounter', August 2012.
Hennessy, Peter, *Having it So Good: Britain in the Fifties*, Penguin Books, 2006.
Hough, Richard, and Denis Richards, *The Battle of Britain*, W. W. Norton, 1990.
Kaplan, Philip and Richard Collier, *Their Finest Hour*, Abbeville Press, 1989.
Kershaw, Alex, *The Few*, De Capo Press, 2006.
Korda, Michael, *With Wings Like Eagles*, Harper Collins e-Books, 2009.
Lloyd, Sir Hugh Pughe, *Briefed to Attack*, London, 1949.
Matheson, Bill, *A Mallee Kid with The Flying Sword*, Self-published, 2004.
Norman, Sir Torquil, *Kick the Tyres, Light the Fires*, Infinite Ideas Limited, 2010.
Pearson, Simon, *The Great Escaper*, Hodder and Stoughton, 2013.
Probert, Henry, *128, The Story of the Royal Air Force Club*, The Royal Air Force Club, 2004.
Richards, Denis, *Royal Air Force, Volume I – The Fight at Odds*, HMSO, 1954.
Richards, Denis and Hilary St. G. Saunders, *Royal Air Force, Vol. II*, HMSO, 1954.
Richey, Paul, *Fighter Pilot*, Published anonymously, 1941.
Saunders Andy, *Battle over Sussex*, Middleton Press, 1989.
Smith, T. Wewege, *Gran Chaco Adventure*, Hutchinson & Co. 1937.
Spears, Sir Edward, *Assignment to Catastrophe*, William Heinemann Ltd, 1954.
Templewood, Lord, *Empire of the Air*, Collins, 1957.
Turner, John Frane, *Fight For The Air*, Naval Institute Press, 2000.
Wellum, Geoffrey, *First Light*, Penguin Books, 2002.

Index